高等职业教育计算机系列教材

软件设计原则与模式
——基于Java/Python语言实现
（微课版）

郭双宙　李　凯　主　编

　　　　　张鼎开　副主编

宋　戈　梁金兰　参　编

电子工业出版社
Publishing House of Electronics Industry
北京·BEIJING

内 容 简 介

软件设计模式分为三大类型：创建型、结构型和行为型。本书共 4 章，分别讲解三大类型中常用的 14 种模式，每种模式都以典型问题为案例，由相应的设计模式给出最佳解决方案。在每章的后半部分都有相应的案例，并分别使用 Java 和 Python 语言实现，以帮助读者快速理解和掌握这些模式，体现教、学、做一体的教学理念。

本书适合作为高等职业院校计算机软件专业的教材，也可供从事软件开发与测试维护的初、中级人员参考。

未经许可，不得以任何方式复制或抄袭本书之部分或全部内容。
版权所有，侵权必究。

图书在版编目（CIP）数据

软件设计原则与模式：基于 Java/Python 语言实现：微课版 / 郭双宙，李凯主编. —北京：电子工业出版社，2022.8
ISBN 978-7-121-43915-5

Ⅰ. ①软… Ⅱ. ①郭… ②李… Ⅲ. ①JAVA 语言－程序设计－高等职业教育－教材 Ⅳ. ①TP312.8

中国版本图书馆 CIP 数据核字（2022）第 118204 号

责任编辑：徐建军　　　　　　　特约编辑：田学清
印　　刷：三河市君旺印务有限公司
装　　订：三河市君旺印务有限公司
出版发行：电子工业出版社
　　　　　北京市海淀区万寿路 173 信箱　　邮编：100036
开　　本：787×1 092　1/16　印张：12.5　字数：336 千字
版　　次：2022 年 8 月第 1 版
印　　次：2022 年 8 月第 1 次印刷
印　　数：1 200 册　　定价：42.00 元

凡所购买电子工业出版社图书有缺损问题，请向购买书店调换。若书店售缺，请与本社发行部联系，联系及邮购电话：(010) 88254888，88258888。
质量投诉请发邮件至 zlts@phei.com.cn，盗版侵权举报请发邮件至 dbqq@phei.com.cn。
本书咨询联系方式：(010) 88254570，xujj@phei.com.cn。

前　言

  编程从某种意义上说是一门"手艺"，因为优雅而高效的代码就如同完美的手工艺品，让人赏心悦目。在雕琢代码的过程中，大有讲究，比如应该用什么架构、哪种模式，还有更多的小细节，比如何时使用异常、怎么给变量起名等。

  真正优秀的代码，是由无数优秀的细节构成的。

  本书要讲解的设计模式其实就是对经常发生的问题提出的解决方案，这种方案经过无数人的测试、使用，千锤百炼之后几乎无懈可击。本书选择的案例都是经过实践、锤炼、选优后的程序，极具参考价值，对学生学习设计模式和 Java/Python 语言都有极大帮助。

  在现实情况中，很多人不使用某种设计模式是因为不了解它的优点，认为这种设计模式很复杂。其实，设计模式的"复杂"就在于它要构造一把"万能钥匙"，提出一种对所有锁适用的开锁方案。对程序员来说，掌握这把"万能钥匙"的好处实在太多了。

  很多程序员在接触设计模式之后，有一种相见恨晚的感觉。有人在学习了设计模式之后，感觉自己好像已经脱胎换骨，达到了新的境界。还有人甚至把是否了解设计模式作为划分程序员编程水平的标准。

  本书共 4 章，分为两部分：第 1 部分（第 1 章）是设计原则简介及必要的 UML 知识；第 2 部分（第 2~4 章）详细介绍了 14 种设计模式，每种设计模式都有一个与之适应的、浅显易懂的例子作为引子，并有详细的 UML 结构设计图及相对应的可运行程序，以帮助读者理解所学的设计模式。

  本书简单易懂，把设计模式的学习门槛降到了最低，使初学者更加容易理解、掌握 14 种设计模式。书中的每个程序都力求简洁明了、易学易用。本书数万行的 Java/Python 代码既可以使读者学习 14 种常用的设计模式，又可以使读者熟悉 Java 和 Python 语言，可谓一举两得。

  本书使用的程序开发工具分别为 MyEclipse 和 PyCharm-Professional。制图软件为 Enterprise Architect，或使用 AmaterasUML 插件在 MyEclipse 中制作结构图。

  本书由郭双宙、李凯担任主编，由张鼎开担任副主编，参加编写工作的还有宋戈、梁金兰。其中，郭双宙负责全书的规划与整体设计，编写全书的主要部分，并负责全书的统稿工作；李凯负责 1.2 节、3.1 节的编写工作，并对本书提供了建设性的建议；张鼎开负责 1.3 节、2.3 节的编写工作；宋戈负责 3.3 节的编写工作；梁金兰负责 1.1 节、4.2 节的编写工作。

  为了方便教师教学，本书配有电子教学课件及相关资源，有此需要的教师可以登录华信教育资源网（www.hxedu.com.cn）免费下载。如果有问题，可以在网站的留言板留言或与电子工业出版社联系（E-mail：hxedu@phei.com.cn）。

教材的编写是一项系统工程，需要在实践中不断完善和改进。同时，由于时间仓促、编者水平有限，书中难免存在疏漏和不足之处，敬请同行专家和广大读者给予批评和指正。

编　者

目 录

第 1 章 软件设计原则与 UML 简介 1

  1.1 开闭原则 2
      1.1.1 应用实例：银行业务 2
      1.1.2 银行业务实现 3
      1.1.3 练习 7
  1.2 里氏替换原则 11
      1.2.1 预备知识 11
      1.2.2 里氏替换原则简介 11
      1.2.3 "乘马说" Java 实现 12
      1.2.4 经典实例：鸵鸟非鸟 13
  1.3 依赖倒置原则 16
      1.3.1 Java 应用实例 16
      1.3.2 依赖倒置原则的意义 16
      1.3.3 依赖倒置原则的优点 17
      1.3.4 依赖倒置原则的例子 17
      1.3.5 练习数据访问 MySQL/Oracle 18
  1.4 接口隔离原则 21
  1.5 合成/聚合复用原则 23
      1.5.1 应用实例 24
      1.5.2 练习 25
  1.6 迪米特法则 29
      1.6.1 迪米特法则简介 29
      1.6.2 违反迪米特法则的设计与实现 30
      1.6.3 遵守迪米特法则的设计与实现 32
  1.7 单一职责原则 34
      1.7.1 应用实例：用户信息管理系统 34
      1.7.2 用户信息管理系统设计与 Java 实现 35
      1.7.3 用户信息管理系统 Python 实现 37
  1.8 UML 简介 38
      1.8.1 依赖 38

	1.8.2	关联	39
	1.8.3	聚合	39
	1.8.4	组合	39
	1.8.5	泛化	40
	1.8.6	实现	40
1.9		设计模式简介	40

第 2 章 创建型模式 42

2.1		简单工厂模式	42
	2.1.1	简单工厂模式的结构	43
	2.1.2	应用系统登录 Java 实现	43
	2.1.3	简单工厂模式的优缺点	46
	2.1.4	练习	46
2.2		工厂方法模式	52
	2.2.1	工厂方法模式的结构	53
	2.2.2	练习	55
2.3		抽象工厂模式	63
	2.3.1	抽象工厂模式的起源	63
	2.3.2	抽象工厂模式的结构	64
	2.3.3	抽象工厂模式的优缺点	67
	2.3.4	练习	67
2.4		单例模式	70
	2.4.1	单例模式的结构	70
	2.4.2	单例模式常见的应用场景	70
	2.4.3	单例模式的类型	71
	2.4.4	练习	75
2.5		多例模式	78
	2.5.1	多例模式结构	78
	2.5.2	练习	79

第 3 章 结构型模式 86

3.1		适配器模式	86
	3.1.1	适配器模式的结构	87
	3.1.2	电源适配器实现	90
	3.1.3	适配器模式的优缺点	92
	3.1.4	练习	93
3.2		默认适配器模式	96
	3.2.1	默认适配器模式的结构	97

3.2.2　练习 99
3.3　装饰模式 102
　　3.3.1　应用实例：孙悟空七十二般变化 102
　　3.3.2　装饰模式的结构 103
　　3.3.3　"孙悟空七十二般变化"Java 实现 105
　　3.3.4　装饰模式的简化 108
　　3.3.5　装饰模式进阶 109
　　3.3.6　练习 110
3.4　门面模式 117
　　3.4.1　什么是门面模式 118
　　3.4.2　门面模式的结构 118
　　3.4.3　门面模式在实际开发中的应用场景 118
　　3.4.4　门面模式进阶 120
　　3.4.5　练习 120

第 4 章　行为型模式 126

4.1　策略模式 126
　　4.1.1　应用实例：旅游出行 126
　　4.1.2　策略模式的结构 127
　　4.1.3　策略模式源代码 127
　　4.1.4　认识策略模式 129
　　4.1.5　策略模式的优缺点 129
　　4.1.6　排序策略系统 Java 实现 130
　　4.1.7　练习 136
4.2　模板方法模式 140
　　4.2.1　模板方法模式的结构 141
　　4.2.2　模板方法模式中的方法 141
　　4.2.3　"西天取经八十一难"Java 实现 143
　　4.2.4　模板方法模式进阶 144
　　4.2.5　练习 146
4.3　命令模式 150
　　4.3.1　命令模式的结构 150
　　4.3.2　应用实例：玉帝宣美猴王上天 153
　　4.3.3　命令模式解析 153
　　4.3.4　命令模式和策略模式的区别 154
　　4.3.5　命令模式的优缺点 155
　　4.3.6　练习 155

4.4 状态模式 164
　　4.4.1 状态模式的结构 165
　　4.4.2 练习 168
4.5 观察者模式 175
　　4.5.1 观察者模式结构 175
　　4.5.2 观察者模式模型 176
　　4.5.3 两种模式的比较 182
　　4.5.4 练习 182

参考文献 192

第 1 章

软件设计原则与 UML 简介

对于每一个程序员来说,要想编写出满足要求的程序,首先要了解软件设计的原则。怎样评价设计的软件?通常有如下要求。

(1)可靠性。

软件的可靠性是指软件在测试运行过程中避免发生故障的能力,以及发生故障后,摆脱和排除故障的能力。

(2)健壮性。

健壮性又称鲁棒性,是指软件能够判断某些输入不符合规范要求,并能有合理的处理方式。软件的健壮性是一个比较模糊的概念,却是非常重要的软件外部衡量标准。

(3)可修改性。

软件的可修改性要求开发者以科学的方法设计软件,使软件有良好的结构和完备的文档,系统性能易于调整。

(4)可理解性。

软件的可理解性是其可靠性和可修改性的前提。它并不仅要求软件的文档清晰可读,而且要求软件本身具有简单明了的结构。

(5)程序简便。

(6)可测试性。

可测试性就是设计一个适当的数据集合,用这个数据集合测试建立的系统,并保证系统得到全面的检验。

(7)效率性。

软件的效率性一般用程序的执行时间和所占用的内存容量来衡量。在达到原理要求的功能指标的前提下,程序运行需要的时间越短、占用存储容量越小,效率越高。

(8)标准化原则。

标准化原则要求软件基于业界的开放式标准,在结构上实现开放,并符合中华人民共和国工业和信息化部的规范。

(9)先进性。

软件的先进性要求该软件满足客户的需求,系统性能可靠,易于维护。

(10)可扩展性。

软件的可扩展性是指软件设计完成后要留有升级接口和升级空间。

满足了软件开发过程中常用的软件设计原则,就容易实现上面的诸多要求。

软件设计的重要性表现在软件的质量上。软件设计描述了软件是如何被分解和集成为组件的,也描述了组件之间的接口以及组件之间是如何发挥软件构建的功能的。

如何设计才能保证软件的质量？这里，我们先给出软件设计的5个基本原则。

（1）要有分层的组织结构，便于对软件各个构件进行控制。

（2）应形成具有独立功能特征的模块（模块化）。

（3）应有性质不同、可区分的数据和过程描述（表达式）。

（4）应尽量降低模块与模块之间、模块与外部环境之间的接口的复杂性。

（5）应利用软件需求分析中得到的信息和可重复的方法。

要想得到一个满意的设计结果，不光要有基本设计原则的指导，还要将系统化的设计方法和科学严格的评审机制相结合。

软件设计原则从宏观上指导着软件设计，但软件设计的具体实现还是要遵循以上软件设计的基本原则。

本书选择了7种常用的软件设计原则，下面介绍7种常用的软件设计原则及在本书中用到的UML图的知识。

1.1 开闭原则

开闭原则（Open-Closed Principle，OCP）是指一个软件实体应当对扩展开放，对修改关闭。

这个原则说的是在设计一个模块的时候，应当使这个模块可以在不被修改的前提下被扩展。换言之，应当可以在不必修改源代码的情况下改变这个模块的行为，在保持系统一定的稳定性的基础上，对系统进行扩展。这是面向对象设计（OOD）的基石，也是最重要的原则。

所有的软件系统有一个共性，即用户对它们的需求会随时间的推移而发生变化。当软件系统面临新的用户需求时，系统的设计必须是稳定的。

西汉扬雄《太玄·玄莹》中有述："夫道有因有循，有革有化。……故知因而不知革，物失其则；知革而不知因，物失其均。"

一个系统因外界变化而出现的问题，就对应上文中的"因"。而系统对扩展开放，就对应上文中的"革"。一个系统在外界变化后不可扩展，则"知因而不知革，物失其则"，即系统失去使用价值；而一个系统动辄需要修改，则"知革而不知因，物失其均"，即找不到导致系统被修改的原因，系统同样会失去使用价值。

这也是对开闭原则早期的阐述之一。

满足开闭原则的设计可以让一个软件系统拥有无可比拟的优越性。

（1）扩展已有的软件系统，可以为软件系统提供新的行为，以满足用户对软件的新需求，使变化中的软件系统有一定的适应性和灵活性。

（2）已有的软件模块，特别是最重要的抽象层模块不能再修改，这就使变化中的软件系统有一定的稳定性和延续性。

具有这两个优点的软件系统是一个在高层次上实现了复用的系统，也是一个易于维护的系统。

1.1.1 应用实例：银行业务

银行业务通常有存款、取款和转账3种，如果不遵守开闭原则，则可能的设计方案如图1-1所示。

第 1 章 软件设计原则与 UML 简介

```
com.开闭原则违反.银行业务.BankingBusiness
• bankSavings(): void
• drawMoney(): void
• transferAccounts(): void
```

图 1-1 违反开闭原则的银行业务设计方案

假如现在业务增加了，比如增加理财功能等，就必须修改 BankingBusiness 业务类。上述设计在有新的需求时必须修改现有代码，像这样需要不断修改的程序，不仅违反开闭原则，也不具有实用价值。

为了达到实用的目的，需要对系统进行抽象化设计：将存款、取款、转账、理财等业务分别抽象为银行业务接口的子类。这样，以后再增加新业务也不用修改现有的程序，只需增加相应子类即可。

修改后的设计方案如图 1-2 所示。

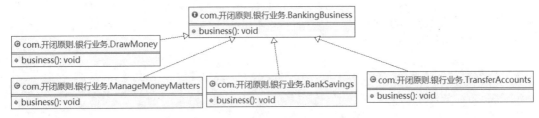

图 1-2 遵守开闭原则的银行业务设计方案

1.1.2 银行业务实现

1. 违反开闭原则的银行业务 Java 实现

```java
package com.开闭原则违反.银行业务;
//客户类
public enum Client {
    存款,取款,转账
}
package com.开闭原则违反.银行业务;
//银行业务类
public class BankingBusiness
{
    public void bankSavings()
    {
        System.out.println("存款");
    }
    public void drawMoney ()
    {
        System.out.println("取款");
    }
    public void transferAccounts()
    {
        System.out.println("转账");
```

```
    }
}
```

测试程序如下：

```java
package com.开闭原则违反.银行业务;
//测试
public class Test {
    public static void main(String[] args){
        Client client = Client.存款;//客户想去银行存款
        BankingBusiness bankingBusiness = new BankingBusiness();

        switch (client){
            case 存款 :bankingBusiness.bankSavings(); break;
            case 取款 :bankingBusiness.drawMoney(); break;
            case 转账 :bankingBusiness.transferAccounts(); break;
            default:break;
        }
    }
}
```

程序运行结果如图 1-3 所示。

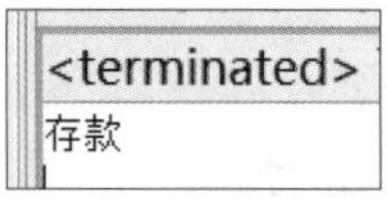

```
<terminated>
存款
```

图 1-3 违反开闭原则的银行业务 Java 实现程序运行结果

2．遵守开闭原则的银行业务 Java 实现

```java
package com.开闭原则.银行业务;
//客户业务种类
public enum Business {
    存款,取款,转账,理财
}
//银行业务
package com.开闭原则.银行业务;
public interface BankingBusiness {
    public void business();
}
//银行业务一：存款
package com.开闭原则.银行业务;
public class BankSavings implements BankingBusiness {
    public void business() {
        // TODO Auto-generated method stub
        System.out.println("银行存款");
    }
}
```

```java
//银行业务二：取款
package com.开闭原则.银行业务;
public class DrawMoney implements BankingBusiness {
    public void business() {
        // TODO Auto-generated method stub
        System.out.println("银行取款");
    }
}

//银行业务三：转账
package com.开闭原则.银行业务;
public class TransferAccounts implements BankingBusiness {
    public void business() {
        // TODO Auto-generated method stub
        System.out.println("银行转账");
    }
}

//银行业务四：理财
package com.开闭原则.银行业务;
public class ManageMoneyMatters implements BankingBusiness {
    public void business() {
        // TODO Auto-generated method stub
        System.out.println("银行理财");
    }
}

//银行客户业务操作
package com.开闭原则.银行业务;
public class Client {
    public static void getBusiness(Business business){
        switch (business) {
            case 存款 :
                (new BankSavings()).business(); break;
            case 取款 :
                (new DrawMoney()).business(); break;
            case 转账 :
                (new TransferAccounts()).business(); break;
            case 理财 :
                (new ManageMoneyMatters()).business(); break;
            default:break;
        }
    }
}
```

测试程序如下：

```java
package com.开闭原则.银行业务;
```

```
//测试
public class Test {
    public static void main(String[] args){
        Business business = Business.转账;//客户想去银行转账
        Client.getBusiness(business);
    }
}
```

程序运行结果如图 1-4 所示。

图 1-4　遵守开闭原则的银行业务 Java 实现程序运行结果

3. 遵守开闭原则的银行业务 Python 实现

```
# 银行业务抽象
import abc
class BankingBusiness(object):
    @abc.abstractmethod
    def business(self):
        pass
# 银行具体业务
class BankSavings(BankingBusiness):
    def business(self):
        print("银行存款")
class DrawMoney(BankingBusiness):
    def business(self):
        print("银行取款")
class TransferAccounts(BankingBusiness):
    def business(self):
        print("银行转账")
class ManageMoneyMatters(BankingBusiness):
    def business(self):
        print("银行理财")
# 银行客户业务
class client(object):
    businessdict = {"存款": BankSavings(), "取款": DrawMoney(),"转账":TransferAccounts(),"理财":ManageMoneyMatters()}
    def getBusiness(business):
        return client.businessdict.get(business).business()
# 测试
if __name__ == '__main__':
```

```
client.getBusiness("取款")
client.getBusiness("理财")
```

程序运行结果如图 1-5 所示。

```
开闭原则.银行业务  ×
C:\Python\Python38-32\python.exe
银行取款
银行理财
```

图 1-5　遵守开闭原则的银行业务 Python 实现程序运行结果

现在的设计简单易懂，条理清晰，对以后的业务扩展也很方便，是非常有竞争力的设计方案。

1.1.3　练习

1．"网购打折" Java 实现

场景描述：现在要出售雪花酥小蛋糕大礼包，雪花酥小蛋糕大礼包的属性有编号、价格、口味。

1）商品原价实现

（1）抽象类（实现公共方法，与打折有关的方法由子类实现）：

```java
package com.开闭原则;
public abstract class Cake {
    private Integer id;
    private String taste;
    private Double price;
    public Cake(Integer id, String taste, Double price) {
        this.id = id;
        this.taste = taste;
        this.price = price;
    }
    public Integer getID() {
        return this.id;
    }
    public String getTaste() {
        return this.taste;
    }
    public Double getPrice() {
        return this.price;
    }
    public abstract String toString();
}
```

（2）实现类：

```java
package com.开闭原则;
public class SnowCake extends Cake {
```

```
    public SnowCake(Integer id, String taste, Double price) {
        super(id, taste, price);
        // TODO Auto-generated constructor stub
    }
    @Override
    public String toString() {
      String str = "SnowCake{" + "id=" + super.getID() + ", taste='" + super.getTaste() + '\'' + ", price=" + super.getPrice() + '}';
        return str;
    }
}
```

2)现在商家要做活动:打 8 折

最佳的方案是通过扩展实现变化,这样不用修改已有的代码。扩展实现的方法是增加一个继承抽象类 Cake 的子类 SnowCakeDiscount,重写 toString()方法。

以后再有其他打折活动也很容易实现,不用修改既有代码,设计方案如图 1-6 所示。

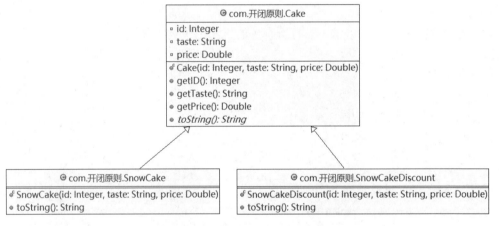

图 1-6 "网购打折"设计方案

子类 SnowCakeDiscount:

```
package com.开闭原则;
//雪花酥小蛋糕打 8 折
public class SnowCakeDiscount extends Cake {
    public SnowCakeDiscount(Integer id, String taste, Double price) {
        super(id, taste, price);
    }
    @Override
    public String toString() {
        return "SnowCake{" +
                "id=" + super.getID() +
                ", taste='" + super.getTaste()+ '\'' +
                ", price=" + super.getPrice()*0.8 +
                '}';
```

        }
    }

测试程序如下：
```java
package com.开闭原则;
public class Test {
    public static void main(String[] args) {
        Cake sc = new SnowCake(100,"原味",55.0);
        System.out.println(sc);
        System.out.println("---------打折后-----------");
        sc = new SnowCakeDiscount(100,"原味",55.0);
        System.out.println(sc);
    }
}
```

程序运行结果如图 1-7 所示。

```
<terminated> Test (3) [Java Application] C:\Java\Genuitec\
SnowCake{id=100, taste='原味', price=55.0}
---------打折后-----------
SnowCake{id=100, taste='原味', price=44.0}
```

图 1-7 "网购打折" 程序运行结果

2. "网购打折" Python 实现

```python
# 网上书店打折销售：所有 50 元以上的书籍 8 折销售，其他书籍 9 折销售
class Book(object):
    def __init__(self, name, price, author):
        self.name = name
        self.price = price
        self.author = author
    def get_name(self):
        pass
    def get_price(self):
        pass
    def get_author(self):
        pass
    def get_book_oldPrice(self):
        pass
    def get_book_newPrice(self):
        pass
class NovelBook(Book):
    def __init__(self, name, price, author):
        super(NovelBook, self).__init__(name, price, author)
    def get_name(self):
        return self.name
    def get_price(self):
```

```python
            return self.price
        def get_author(self):
            return self.author
        def get_book_oldPrice(self):
            return "Book name: " + self.get_name() + " Book author: " + self.get_author() +\
                " Book price: " + str(self.get_price() / 100.0) + "元"
        def get_book_newPrice(self):
            return "Book name: " + self.get_name() + " Book author: " + self.get_author() +\
                " Book price: " + str(self.get_price() / 100.0) + "元"
    class OffNovelBook(NovelBook):
    # 新增子类用于扩展
        def __init__(self, name, price, author):
            super(OffNovelBook, self).__init__(name, price, author)
        def get_name(self):
            return self.name
        def get_price(self):
            origin_price = super(OffNovelBook, self).get_price()
            off_price = 0
            if origin_price >= 5000:
                off_price = origin_price * 0.8
            else:
                off_price = origin_price * 0.9
            return off_price
        def get_author(self):
            return self.author
        def get_book_oldPrice(self):
            off_book_info = super(OffNovelBook, self).get_book_oldPrice()
            return off_book_info
    class BookStore(object):
        def __init__(self):
            self.book_list = []
            self.book_list.append(OffNovelBook("西游记", 3000, "吴承恩"))
            self.book_list.append(OffNovelBook("三国演义", 6000, "罗贯中"))
            self.book_list.append(OffNovelBook("红楼梦", 8000, "曹雪芹"))
        def sell(self, book):
            print("Sell Info: {}".format(book.get_book_oldPrice()))
            print("Sell Info: {}".format(book.get_book_oldPrice()))
    if __name__ == '__main__':
        book_store = BookStore()
        for book in book_store.book_list:
            book_store.sell(book)
```

1.2 里氏替换原则

《墨子·小取》中有述:"白马,马也;乘白马,乘马也。骊马(黑马),马也;乘骊马,乘马也。"我们用这个例子理解里氏替换原则(Liskov Substitution Principle,LSP)。

1.2.1 预备知识

在开始学习里氏替换原则前,有必要先提一下继承(Inheritance)。因为里氏替换原则是一条非常具体的,和类继承有关的原则。

在面向对象程序设计的世界里,继承是一个非常特殊的存在,它像一把无坚不摧的双刃剑,强大却危险。合理使用继承,可以大大减少类与类之间的重复代码,让程序事半功倍。而不合理的继承关系,则会让类与类之间建立错误的强耦合,产生大段难以理解和维护的代码。

因此,人们对继承的态度也可以大致分为两类。大多数人认为,继承和多态、封装等特性一样,属于面向对象编程的核心特征。而另一部分人认为,继承带来的坏处远比好处多。

在实际工作中,继承确实极易被误用。设计出合理的继承关系是一件需要深思熟虑的困难事。不过幸运的是,在这方面,里氏替换原则为我们提供了非常好的指导意义。

在本节中的实现之一"乘马说"的"马"就是一种抽象角色。

1.2.2 里氏替换原则简介

里氏替换原则是继承复用的基石。里氏替换原则的意思是:在程序中如果使用一个父类,那么一定可以使用其子类替换,而且程序根本不能察觉出父类对象和子类对象的区别。只有子类可以替换父类,软件单位的功能才能不受影响,父类才能真正被复用,而子类才能在父类的基础上增加新功能。

反过来的替换不成立,即父类不能替换子类。如"乘马说"的"马"作为抽象角色就不能代替所有继承"马"的子类。

注意应当尽量从抽象类继承,而不从具体类继承。一般而言,如果两个具体类 A 和 B 有继承关系,那么一个最简单的修改方案是先建立一个抽象类 C,然后让类 A 和 B 成为抽象类 C 的子类。即如果有一个由继承关系形成的等级结构,那么在等级结构的树状图上,所有的树叶节点应当是具体类,而所有的树枝节点应当是抽象类或者接口,如图 1-8 所示。

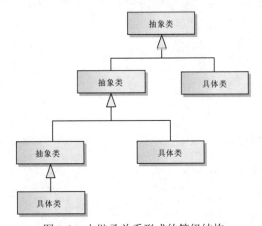

图 1-8 由继承关系形成的等级结构

1.2.3 "乘马说" Java 实现

《墨子·小取》中"乘马说"的结构如图 1-9 所示。

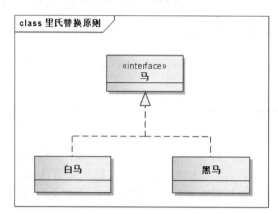

图 1-9 "乘马说"的结构

对骑行者来说，骑白马类的对象白马和骑黑马类的对象黑马的结果是一样的，都能骑马到达目的地。这就是里氏替换原则的一个实际应用。

代码实现如下：

```java
package com.里氏替换原则;
public interface 马 {
    void 骑马();
}
package com.里氏替换原则;
public class 白马 implements 马 {

    public void 骑马() {
        System.out.println("我属于白马!");
    }

}
package com.里氏替换原则;
public class 黑马 implements 马 {

    public void 骑马() {
        System.out.println("我属于黑马!");
    }

}
```

测试程序如下：

```java
package com.里氏替换原则;
public class Test {
    public static void main(String[] args) {
        // TODO Auto-generated method stub
        马 白龙马 = new 白马();
```

```
        白龙马.骑马();

        马 小黑马 = new 黑马();
        小黑马.骑马();
    }
}
```

1.2.4 经典实例:鸵鸟非鸟

"鸵鸟非鸟"是理解里氏替换原则的经典例子之一。从生物学的角度来看,鸵鸟是一种鸟,但是这种鸟不会飞。假设现在有这样一个需求:计算鸟飞越黄河所需的时间。我们用麻雀和鸵鸟两种鸟来进行比较。

1. 违反里氏替换原则的"鸵鸟非鸟"Java 实现

违反里氏替换原则的"鸵鸟非鸟"设计方案如图 1-10 所示。

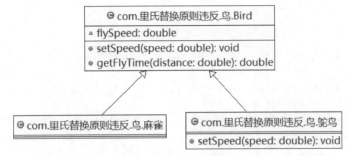

图 1-10 违反里氏替换原则的"鸵鸟非鸟"设计方案

```java
//父类
package com.里氏替换原则违反.鸟;

public abstract class Bird {
    double flySpeed;
    public void setSpeed(double speed) {
        flySpeed = speed;
    }
    public double getFlyTime(double distance) {
        return (distance / flySpeed*60);
    }
}
//子类一
package com.里氏替换原则违反.鸟;
public class 鸵鸟 extends Bird{
    public void setSpeed(double speed) {
        flySpeed = 0;
    }
}
//子类二
```

```java
package com.里氏替换原则违反.鸟;
public class 麻雀 extends Bird {

}
```

测试程序如下：

```java
package com.里氏替换原则违反.鸟;
public class Test {
    /**
     * @param args
     */
    public static void main(String[] args) {
        // TODO Auto-generated method stub
        Bird 小麻雀 = new 麻雀();
        小麻雀.setSpeed(120);
        Bird 大鸵鸟 = new 鸵鸟();
        大鸵鸟.setSpeed(120);
        try {
            System.out.println("麻雀以： " + 小麻雀.getFlyTime(6) + " 小时飞越黄河！");
            System.out.println("鸵鸟以： " + 大鸵鸟.getFlyTime(6) + " 小时飞越黄河！");
        }
        catch (Exception err) {
            System.out.println(err);
        }
    }
}
```

如果使用麻雀来测试这段代码，没有问题，结果正确，符合预期，系统输出了麻雀飞越黄河需要的时间。如果我们拿鸵鸟测试这段代码，结果代码发生了系统除零的异常，明显不符合我们的预期。测试程序运行结果如图 1-11 所示。

```
<terminated> Test (4) [Java Applicat
麻雀以：  3.0  小时飞越黄河！
鸵鸟以：  Infinity  小时飞越黄河！
```

图 1-11 "鸵鸟非鸟"程序运行结果

这是因为鸵鸟类和 Bird 类之间的继承关系违反了里氏替换原则，它们之间的继承关系不成立。

鸵鸟到底是不是鸟？鸵鸟是鸟，也不是鸟，这个结论似乎是一个悖论。产生这种混乱有两方面的原因。

（1）没有搞清楚类的继承关系的定义。

在面向对象的设计中，类的继承关注的是对象的行为，它是使用"行为"对对象进行分类的，只有行为一致的对象才能抽象出一个类。类的继承关系就是一种"Is-A"关系，这种"Is-A"关系实际上指的是行为上的"Is-A"关系，我们可以把它描述为"Act-As"。

（2）设计依赖于用户要求和具体环境。

继承关系要求子类具有基类全部的行为。这里的行为是指落在需求范围内的行为。

- A 需求期望鸟类提供与飞翔有关的行为，而鸵鸟在飞翔这一点上无法从鸟类派生。
- B 需求期望鸟类提供与羽毛有关的行为，而鸵鸟具备了鸟类全部的行为特征，鸵鸟类就能够从鸟类派生。

所有派生类的行为和功能必须和使用者对其基类的期望保持一致，如果派生类无法实现这一点，那么必然违反里氏替换原则。里氏替换原则就是在设计时要避免出现派生类与基类不一致的行为，此时，修改各种代码也会违反开闭原则。

增加两个抽象类分别作为"会飞的鸟"和"不会飞的鸟"的父类，这样就完美解决了类型不同带来的烦扰，这时的代码也符合了开闭原则。Java 代码除了增加不会飞的鸟和会飞的鸟两个抽象类，其他代码与上文相近，读者可自行实现。

这时，"鸵鸟非鸟"设计方案如 1-12 所示。

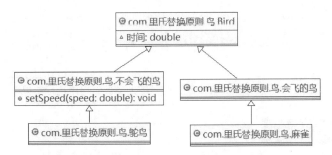

图 1-12 遵守里氏替换原则的"鸵鸟非鸟"设计方案

2. 遵守里氏替换原则的"鸵鸟非鸟"Python 实现

Python 实现代码如下：

```python
class Bird(object):
    def __init__(self, flySpeed):
        self.flySpeed = flySpeed

    def setSpeed(self, speed):
        self.flySpeed = speed

    def getFlyTime(self, distance):
        return distance / self.flySpeed * 60
class 会飞的鸟(Bird):
    pass

class 不会飞的鸟(Bird):
    def setSpeed(self, speed):
        self.flySpeed = 0

class 麻雀(会飞的鸟):
    pass

class 鸵鸟(不会飞的鸟):
    pass
if __name__ == '__main__':
    小麻雀 = 麻雀(0)
    小麻雀.setSpeed(120)
```

```
    print("麻雀以： "，小麻雀.getFlyTime(6) ，" 小时飞越黄河！")
大鸵鸟 = 鸵鸟(0)
大鸵鸟.setSpeed(120)
    print("鸵鸟以： "，大鸵鸟.getFlyTime(6) ，" 小时飞越黄河！")
```

Python实现的程序运行结果如图1-13所示。

```
里氏替换原则的"鸵鸟非鸟"
C:\Python\Python38-32\python.exe E:/教学资料/教学/设计模式2019最新版/Python实现/抽象类实例/venv/设计模式/里氏替换原则的"鸵鸟非鸟".py
Traceback (most recent call last):
麻雀以： 3.0 小时飞越黄河！
  File "E:/教学资料/教学/设计模式2019最新版/Python实现/抽象类实例/venv/设计模式/里氏替换原则的"鸵鸟非鸟".py", line 30, in <module>
    print("鸵鸟以： ，大鸵鸟.getFlyTime(6) ，" 小时飞越黄河！")
  File "E:/教学资料/教学/设计模式2019最新版/Python实现/抽象类实例/venv/设计模式/里氏替换原则的"鸵鸟非鸟".py", line 9, in getFlyTime
    return distance / self.flySpeed * 60
ZeroDivisionError: division by zero
```

图1-13 遵守里氏替换原则的"鸵鸟非鸟"程序运行结果

1.3 依赖倒置原则

依赖倒置原则（Dependence Inversion Principle，DIP）要求客户端依赖于抽象耦合。

依赖倒置原则表述一：抽象不应当依赖于细节，细节应当依赖于抽象。

依赖倒置原则表述二：要针对接口编程，不要针对实现编程。

1.3.1 Java应用实例

（1）private List list = new ArrayList(); //List 是接口

（2）public List method(List sample){

……

}

不要针对实现编程，意思是不应当使用具体类进行变量的类型声明、参量的类型声明、方法的返还类型声明等。例如：

（1）private ArrayList list = new ArrayList();//ArrayList 是 List 的一个子类

（2）public ArrayList method(ArrayList sample){

……

}

依赖的3种实现：

（1）依赖对象通过构造函数以参数形式传递。

（2）依赖对象通过 Setter 方法以参数形式传递。

（3）依赖对象通过接口或抽象类继承的方式声明。

1.3.2 依赖倒置原则的意义

只要一个被引用的对象存在抽象类型，就应当在任何引用此对象的位置使用抽象类型，包括参量的类型声明、方法返还的类型声明、属性变量的类型声明等。

依赖倒置原则是七大原则中最难实现的原则，它是实现开闭原则的重要途径，没有实现依赖倒置原则，就无法实现开闭原则。在项目中只要记住"面向接口编程"，就基本上抓住了依赖倒置原则的核心。

1.3.3 依赖倒置原则的优点

可以通过抽象使各个类或模块的实现独立，不互相影响，实现模块间的松耦合，这也是依赖倒置原则的本质。

依赖倒置原则可以规避一些非技术因素引起的问题。项目越大，需求变化的概率越大，通过采用依赖倒置原则设计的接口或抽象类对实现类进行约束，可以减少由需求变化引起的工作量剧增的情况。同时，如果发生人员变动，只要文档完善，就可以让维护人员轻松地扩展和维护。

依赖倒置原则可以促进并行开发。例如，两个类之间有依赖关系，规范已经定好了，只要制定两者之间的接口（或抽象类）就可以独立开发了，而且项目之间的单元测试也可以独立地运行。测试驱动的开发更是依赖倒置原则的最高级应用（特别适合负责项目的人员整体水平较低时使用）。

1.3.4 依赖倒置原则的例子

高层模块不应该依赖于低层模块，二者都应该依赖于抽象。依赖倒置原则在分层架构模式中，得到了淋漓尽致的运用。

例如，业务逻辑操作层（高层模块）的对象不应该直接依赖于数据访问层（低层模块）的具体实现对象，而应该通过数据访问层的抽象接口进行访问。具体来说就是，我们在编写业务代码的时候，不应该在某个业务代码处直接用 JDBC 操作数据库，业务层方法的参数应该是数据访问层的抽象，如图 1-14 所示。

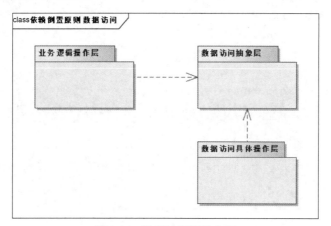

图 1-14 数据访问设计方案

如果高层模块直接依赖于低层模块，一旦低层模块发生变化，就会影响到高层模块。在引入抽象之后，对高层模块而言，低层模块的实现是可替换的。这实际上也是开闭原则的体现。

这一原则同时体现了软件设计对"间接"的追求。图 1-14 中的数据访问抽象层就是在设计中引入的间接访问层。

1.3.5 练习数据访问 MySQL/Oracle

1. 违反依赖倒置原则的"数据访问"Java 实现

上面提到了违反依赖倒置原则的后果:如果高层模块直接依赖于低层模块,一旦低层模块发生变化,就会影响到高层模块。下面就是违反依赖倒置原则的设计方案,如图 1-15 所示。

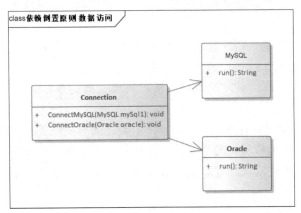

图 1-15 违反依赖倒置原则的"数据访问"设计方案

代码实现如下。

连接各数据库:

```java
package com.依赖倒置原则违反.连接数据库;
//连接各数据库
public class Connection {
    private String name;

    public Connection(String name) {
        this.name = name;
    }

    public void connDatabase(MySQL mysql){
        System.out.println(this.name + "连接成功," + mysql.run());
    }
    public void connDatabase(Oracle oracle){
        System.out.println(this.name + "连接成功," + oracle.run());
    }
}
```

MySQL:

```java
package com.依赖倒置原则违反.连接数据库;
public class MySQL {
    public String run(){
        return "这里编写连接MySQL代码! ";
    }
}
```

Oracle：

```
package com.依赖倒置原则违反.连接数据库;
public class Oracle {
    public String run(){
        return "这里编写连接Oracle代码! ";
    }
}
```

测试程序如下：

```
package com.依赖倒置原则违反.连接数据库;
public class Client {
    public static void main(String[] args) {
        Connection conn = new Connection("MySQL");
        MySQL mysql = new MySQL();
        conn.connDatabase(mysql);
        conn = new Connection("Oracle");
        Oracle oracle = new Oracle();
        conn.connDatabase(oracle);
    }
}
```

程序运行结果如图 1-16 所示。

由图 1-16 可知，每增加一种数据库就要修改 Connection 类，增加一个对应的 Conn×××()方法，造成这种现象的原因是设计出了问题：Connection 类和 MySQL/Oracle 类之间是紧耦合的关系，导致系统的可维护性降低，可读性降低。既不符合开闭原则、里氏替换原则，也不符合依赖倒置原则。

图 1-16 "数据访问"程序运行结果

遵守依赖倒置原则的"数据访问"设计方案如图 1-17 所示。

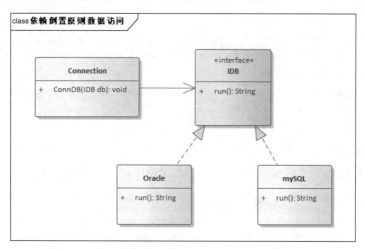

图 1-17 遵守依赖倒置原则的"数据访问"设计方案

从设计图中可以看出,在现在的设计中,底层程序 Connection 类已经和具体的数据库无关,这样就可以根据需求扩展和连接不同的数据库。Java 代码可参考 Python 实例。

2. 遵守依赖倒置原则的"数据访问"Python 实现

```python
class Connection(object):
    def __init__(self,db):
        self.db = db

    def ConnDB(self,db):
        self.db = db
        print(db.name , "连接成功," ,db.run());

class IDB(object):
    def run(self):
        pass

class MySQL(IDB):
    def __init__(self,name):
        self.name = name
    def run(self):
        return "这里编写连接MySQL代码!"

class Oracle(IDB):
    def __init__(self,name):
        self.name = name
    def run(self):
        return "这里编写连接Oracle代码!"

if __name__ == '__main__':
    def connDB(db):
        conn = Connection(db)
        conn.ConnDB(db)

    mydb = MySQL("MySQL")
    connDB(mydb)

    mydb = Oracle("Oracle")
    connDB(mydb)
```

遵守依赖倒置原则的"数据访问"Python 实现程序运行结果如图 1-18 所示。

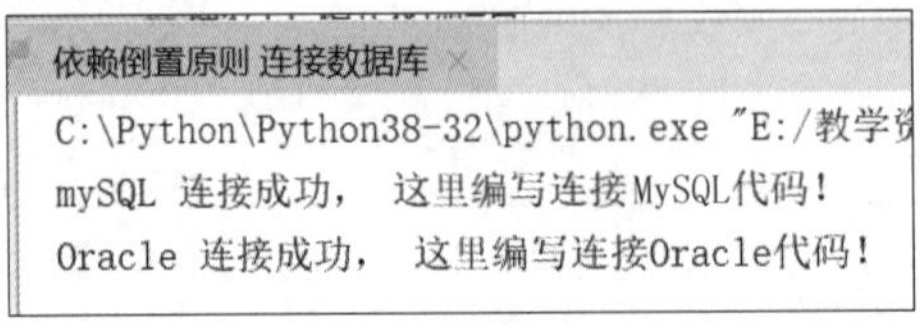

```
依赖倒置原则 连接数据库  ×
C:\Python\Python38-32\python.exe "E:/教学资
mySQL 连接成功,  这里编写连接MySQL代码!
Oracle 连接成功,  这里编写连接Oracle代码!
```

图 1-18 遵守依赖倒置原则的"数据访问"Python 实现程序运行结果

1.4 接口隔离原则

接口隔离原则(Interface Segregation Principle,ISP)有以下要求。
(1)使用多个专门的接口比使用单一的总接口要好。
(2)客户端不应该依赖它不需要的接口。
(3)类间的依赖关系应该建立在最小的接口上。

胖接口会导致客户程序之间产生不正常的并且有害的耦合关系。当一个客户程序要求该胖接口进行一个改动时,会影响到其他的客户程序。因此,客户程序应该仅仅依赖它们实际需要调用的方法。

接口隔离原则的使用场合:提供调用者需要的方法,屏蔽不需要的方法。

应用实例:电子商务系统订单类 Java 实现

1. 需求分析

任何一个电子商务系统都会有订单这个类,通常有 3 个地方会使用订单类。
(1)门户:只能有查询订单方法。
(2)外部系统:只有添加和查询订单方法。
(3)管理后台:添加、删除、修改、查询订单方法都要用到。

根据接口隔离原则,一个类对另外一个类的依赖关系应当建立在最小的接口上。例如,对于门户,它只能依赖有一个查询方法的接口,而 admin 用户必须有增、删、改、查功能。

该系统的设计思路:先定义 3 个单一接口分别完成各自的任务,再定义一个订单类继承这 3 个接口,其结构如图 1-19 所示。

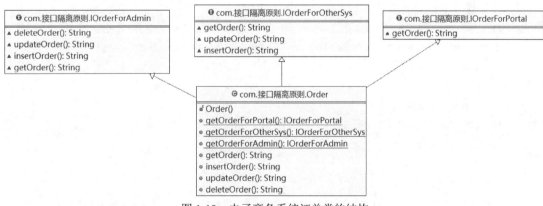

图 1-19 电子商务系统订单类的结构

2. 系统代码

门户接口 IOrderForPortal:

```
package com.接口隔离原则;
interface IOrderForPortal{
    String getOrder();
}
```

外部系统接口：

```java
package com.接口隔离原则;
public interface IOrderForOtherSys {
    String getOrder();
String updateOrder();
String insertOrder();
}
```

管理后台接口：

```java
package com.接口隔离原则;
interface IOrderForAdmin{
  String deleteOrder();
  String updateOrder();
  String insertOrder();
  String getOrder();
}
```

实现接口的订单类 Order：

```java
package com.接口隔离原则;
public class Order implements IOrderForPortal,IOrderForOtherSys,IOrderForAdmin{
  private Order(){
  /**
   * 什么都不做，就是为了不让直接生成一个新的对象，防止客户端直接生成一个新的对象，然后访问它不需要的方法
   */
  }
  //返回给 Portal
  public static IOrderForPortal getOrderForPortal(){
     return (IOrderForPortal)new Order();
  }
  //返回给 OtherSys
  public static IOrderForOtherSys getOrderForOtherSys(){
    return (IOrderForOtherSys)new Order();
  }
  //返回给 Admin
  public static IOrderForAdmin getOrderForAdmin(){
    return (IOrderForAdmin)new Order();
  }
  //下面是接口方法的实现，只返回了一个 String，用于演示
  public String getOrder(){
    return "getOrder()方法";
  }
  public String insertOrder(){
    return "insertOrder()方法";
```

```
    }
    public String updateOrder(){
      return "updateOrder()方法";
    }
    public String deleteOrder(){
      return "deleteOrder()方法";
    }
}
```

测试程序如下:

```
package com.接口隔离原则;
public class Client {
    public static void main(String[] args){
    IOrderForPortal orderForPortal =Order.getOrderForPortal();
    IOrderForOtherSys orderForOtherSys =Order.getOrderForOtherSys();
    IOrderForAdmin orderForAdmin = Order.getOrderForAdmin();

    System.out.println("门户系统可调用:\t" + orderForPortal.getOrder());
    System.out.println("外部系统可调用:\t"+orderForOtherSys.getOrder()+";"+orderForOtherSys.insertOrder()+";"+orderForOtherSys.updateOrder());
    System.out.println("管理后台可调用:\t"+orderForAdmin.getOrder()+";"+orderForAdmin.insertOrder()+";"+orderForAdmin.updateOrder()+";"+orderForAdmin.deleteOrder());
    }
}
```

3. 运行结果

程序的运行结果如图 1-20 所示。

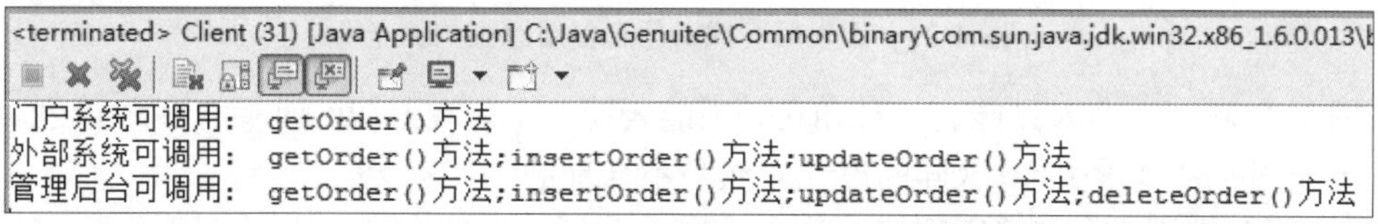

图 1-20 电子商务系统订单类程序运行结果

这样就能很好地满足接口隔离原则了，调用者只能访问自己的方法，不能访问其他的方法。

1.5 合成 / 聚合复用原则

合成 / 聚合复用原则（Composite/Aggregate Reuse Principle，CARP）要求尽量使用合成 / 聚合，尽量不使用继承。

合成和聚合都是对象建模中关联（Association）关系的一种。

合成和聚合的区别如下。

（1）合成是一种更强的聚合，部分组成整体，而且不可分割，部分不能脱离整体而单独存在。在合成关系中，部分和整体的生命周期一样，合成的新的对象完全支配其组成部分，包括它们的创建和销毁。一个合成关系中的成分对象是不能与另一个合成关系共享的。

（2）聚合表示整体与部分的关系，表示"含有"。整体由部分组合而成，部分可以脱离整体作为独立的个体存在。

只有在满足以下全部条件时，才使用继承关系。

（1）子类是超类的一个特殊种类，而不是超类的一个角色，也就是区分"Has-A"和"Is-A"。只有"Is-A"关系才符合继承关系，"Has-A"关系应当使用聚合描述。

（2）永远不会出现需要将子类转换为另一个类的子类的情况。如果不能肯定一个子类将来是否会转换为另一个类的子类，就不要使用继承。

（3）子类具有扩展超类的责任，而不具有置换或注销超类的责任。如果一个子类需要大量地置换超类的行为，那么这个类就不应该是这个超类的子类。

错误地使用继承而不是合成/聚合的一个常见原因是把"Has-A"当成了"Is-A"。"Is-A"代表一个类是另一个类的子类。而"Has-A"代表一个类是另一个类的一个角色，而不是子类。

1.5.1 应用实例

1. 不使用合成/聚合复用原则

错误地使用继承而不使用合成/聚合的一个常见的原因是把"Has-A"当作"Is-A"，如图 1-21 所示。

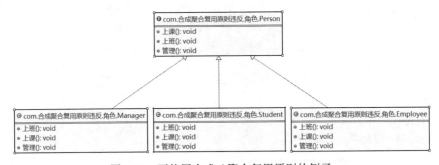

图 1-21　不使用合成/聚合复用原则的例子

在图 1-21 中，一个人只能有一种角色，不能兼职。一个人无法同时拥有多种角色，是"雇员"就不能同时是"学生"，这样的设计显然是不合理的。

2. 使用合成/聚合复用原则

实际上，"经理""雇员""学生"描述的都是一种角色，比如一个人可以是"经理""雇员"，也可以是"学生"。

不使用合成/聚合复用原则的例子的错误在于把"角色"的等级结构与"人"的等级结构混淆了，误把"Has-A"当作"Is-A"，修正后的结构如图 1-22 所示。

图 1-22　使用合成/聚合复用原则的例子

现在编程就灵活多了，把 Role 作为人的属性，同一个人通过调用不同子类就可以很方便地拥有不同角色。

1.5.2　练习

1. 违反合成 / 聚合复用原则的 Java 实现

```java
package com.合成聚合复用原则违反.角色;
public interface Person {
    public void 上课();
    public void 上班();
    public void 管理();
}
package com.合成聚合复用原则违反.角色;
public class Student implements Person {
    public void 上班() {
        // TODO Auto-generated method stub
        throw new RuntimeException("I'm not an Employee!" );
    }
    public void 上课() {
        // TODO Auto-generated method stub
        System.out.println("I'm a Student!");
    }
    public void 管理() {
        // TODO Auto-generated method stub
        throw new RuntimeException("I'm not a Manager!");
    }
}
package com.合成聚合复用原则违反.角色;
public class Manager implements Person {
    public void 上班() {
        // TODO Auto-generated method stub
        throw new RuntimeException("I'm not an Employee!" );
    }
    public void 上课() {
        // TODO Auto-generated method stub
        throw new RuntimeException("I'm not a Student!" );
    }
    public void 管理() {
        // TODO Auto-generated method stub
        System.out.println("I'm a Manager!");
    }
}
package com.合成聚合复用原则违反.角色;
```

```java
public class Employee implements Person {
    public void 上班() {
        // TODO Auto-generated method stub
        System.out.println("I'm an Employee!");
    }
    public void 上课() {
        // TODO Auto-generated method stub
        throw new RuntimeException("I'm not a Student!" ); }
    public void 管理() {
        // TODO Auto-generated method stub
        throw new RuntimeException("I'm not a Manager!");
    }
}
```

测试程序如下：

```java
package com.合成聚合复用原则违反.角色;
public class Test {
    /**
     * @param args
     */
    public static void main(String[] args) {
        // TODO Auto-generated method stub
        Person p = new Student();
        p.上课();
        p.上班();
    }
}
```

运行结果如图 1-23 所示。

```
<terminated> Test (6) [Java Application] C:\Java\Genuitec\Common\binary\com.sur
Exception in thread "main" I'm a Student!
java.lang.RuntimeException: I'm not an Employee!
        at com.合成聚合复用原则违反.角色.Student.上班(Student.java:7)
        at com.合成聚合复用原则违反.角色.Test.main(Test.java:13)
```

图 1-23　违反合成 / 聚合复用原则的程序运行结果

2. 遵守合成 / 聚合复用原则的 Python 实现

```python
# 父类
from abc import abstractmethod, ABCMeta
class Role(metaclass=ABCMeta):
    def doThing(self):
        @abstractmethod
        def doThing(self):
            pass
# 子类
```

```python
class Manager(Role):
    def doThing(self):
        print("I'm a Manager!")
# 子类
class Employee(Role):
    def doThing(self):
        print("I'm an Employee!")
# 子类
class Student(Role):
    def doThing(self):
        print("I'm a Student!")
# 调用程序
class Person():
role =Role()
def __init__(self,role):
    self.role = role
def work(self,role):
    self.role.doThing()

# 测试程序如下：
if __name__ == '__main__':
    manager = Manager()
    employee = Employee()
    student = Student()

    person = Person(manager)
    person.work(person.__class__.__name__)

    person = Person(employee)
    person.work(person.__class__.__name__)

    person = Person(student)
    person.work(person.__class__.__name__)
```

程序运行结果如图 1-24 所示。

图 1-24　遵守合成 / 聚合复用原则的 Python 实现程序运行结果

从运行结果可知，一个人可以为不同角色。

3. 遵守合成/聚合复用原则的 Java 实现

```java
package com.合成聚合复用原则.角色;
public abstract class Role {
    public abstract void doThing(Role role);
}
package com.合成聚合复用原则.角色;
public class Student extends Role {
    private void 上课() {
        // TODO Auto-generated method stub
        System.out.println("I'm a Student!");
    }
    public void doThing(Role role){
        this.上课();
    }
}
package com.合成聚合复用原则.角色;
public class Employee extends Role {
    private void 上班() {
        System.out.println("I'm an Employee!");
    }
    public void doThing(Role role){
        this.上班();
    }
}
package com.合成聚合复用原则.角色;
public class Manager extends Role {
    private void 管理() {
        System.out.println("I'm a Manager!");;
    }
    public void doThing(Role role){
        this.管理();
    }
}
package com.合成聚合复用原则.角色;
public class Person {
    public void work(Role role){
        role.doThing(role);
    }
}
```

测试程序如下:

```java
package com.合成聚合复用原则.角色;
public class Test {
```

```java
    /**
     * @param args
     */
    public static void main(String[] args) {
        Role student = new Student();
        Role employee = new Employee();
        Role manager = new Manager();
        Person p = new Person();
        p.work(student);
        p.work(employee);
        p.work(manager);
    }
}
```

程序运行结果如图 1-25 所示。

```
<terminated> Test (7) [Java Application]
I'm a Student!
I'm an Employee!
I'm a Manager!
```

图 1-25　遵守合成 / 聚合复用原则的 Java 实现程序运行结果

从运行结果可知，和 Python 实现的程序相同，一个人可以为不同角色。

1.6　迪米特法则

我们都知道软件结构设计的总原则是"低耦合，高内聚"。无论是面向过程编程还是面向对象编程，只有使各个模块之间的耦合尽量低，才能提高代码的复用率。低耦合的优点不言而喻，但是，什么样的编程才能做到低耦合呢？这正是迪米特法则要完成的。

1.6.1　迪米特法则简介

迪米特法则（Law of Demeter，LoD）又被称为最少知识原则（The Least Knowledge Principle，LKP），就是说，一个对象应当尽可能少地了解其他对象。

迪米特法则的核心观念就是类间解耦。只有实现低耦合，才能提高类的复用率。

迪米特法则有以下几种表述方式。

（1）只与你直接的朋友们通信（Only talk to your immediate friends）。

（2）不要和"陌生人"说话（Don't talk to strangers）。

（3）每一个软件单位只有其他单位最少的知识，而且局限于那些本单位密切相关的软件单位。

就是说，如果两个类不必直接通信，那么这两个类就不应当发生直接的相互作用，如果其中的一个类需要调用另一个类的某一个方法，可以通过第三者转发这个调用。

1.6.2 违反迪米特法则的设计与实现

我们在安装软件时，经常有一个导向动作，第一步是确认是否安装，第二步是确认许可，然后选择安装目录。这是一个典型的顺序执行动作，具体到程序中就是调用一个或者多个类，先执行第一个方法，再执行是第二个方法，根据返回结果再来看是否调用第三个方法。违反迪米特法则的设计方案如图1-26所示。

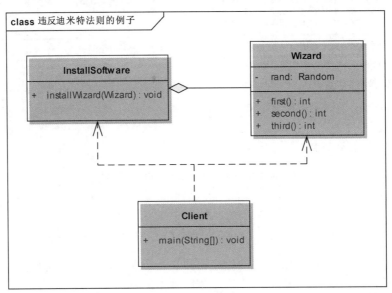

图1-26 违反迪米特法则的设计方案

这是一个很简单的类图，此类图实现软件安装的过程，其中first方法定义第一步做什么，second方法定义第二步做什么，违反迪米特法则的代码如下。

Wizard 类：

```java
package com.no迪米特法则;
import java.util.Random;
public class Wizard {
    private Random rand=new Random(System.currentTimeMillis());
    //第一步
    public int first(){
        System.out.println("执行第一个方法...");
        return rand.nextInt(1010);
    }
    //第二步
    public int second(){
        System.out.println("执行第二个方法...");
        return rand.nextInt(100);
    }
    //第三个方法
    public int third(){
```

```
        System.out.println("执行第三个方法...");
        return rand.nextInt(100);
    }
}
```

InstallSoftware 类：

```
package com.no迪米特法则;
public class InstallSoftware {
    public void installWizard(Wizard wizard){
        int first=wizard.first();
        System.out.println("first返回的结果是： " + first);
        //根据first返回的结果,看是否需要执行second
        if(first>50){
            int second=wizard.second();
            System.out.println("second返回的结果是： " + second);
            if(second>50){
                int third=wizard.third();
                System.out.println("third返回的结果是： " + third);
                if(third >50){
                    wizard.first();
                    System.out.println("third返回的结果>50,继续运行wizard.first();");
                }else
                    System.out.println("因third返回的结果<50,不再执行下一步操作。");
            }else
                System.out.println("因second返回的结果<50,不再执行下一步操作。");
        }else
            System.out.println("因first返回的结果<50,不再执行下一步操作。");
    }
}
```

测试程序如下：

```
package com.no迪米特法则;
public class Client {
    public static void main(String[] args) {
        Wizard wizard = new Wizard();
        InstallSoftware IS = new InstallSoftware();
        IS.installWizard(wizard);
    }
}
```

虽然程序简单，但是隐藏的问题可不简单。Wizard 类把太多的方法暴露给 InstallSoftware 类，两者的朋友关系太紧密了，耦合关系变得异常牢固。如果要将 Wizard 类中 first 方法返回值的类型由 int 改为 boolean，就需要修改 InstallSoftware 类，从而让修改的风险扩散。

程序运行结果如图 1-27 所示。

图 1-27　程序运行结果

1.6.3　遵守迪米特法则的设计与实现

前例的耦合是极度不合适的，我们需要对设计进行重构，重构后的类图如图 1-28 所示。

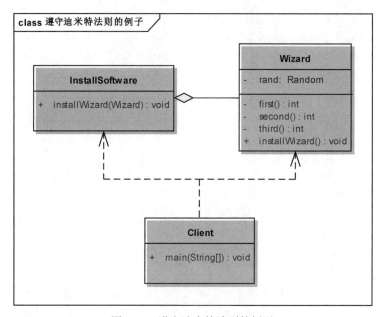

图 1-28　遵守迪米特法则的例子

注意：遵守迪米特法则的 Wizard 类中的方法与前面的不同，增加了 installWizard()。代码中的下画线部分显示了与前面的区别。

```java
package com.迪米特法则;
import java.util.Random;
public class Wizard {
    private Random rand=new Random(System.currentTimeMillis());
    //第一个方法
    private int first(){
        System.out.println("执行第一个方法...");
        return rand.nextInt(1010);
    }
    //第二个方法
    private int second(){
        System.out.println("执行第二个方法...");
```

```java
        return rand.nextInt(100);
    }
    //第三个方法
    private int  third(){
      System.out.println("执行第三个方法...");
    return rand.nextInt(100);
    }
    //增加了一个软件安装方法
    public void  installWizard(){
      int first=this.first();
      System.out.println("first 返回的结果是: " + first);
        //根据first返回的结果,看是否需要执行second
        if(first>50){
            int second=this.second();
            System.out.println("second 返回的结果是: " + second);
            if(second>50){
            int third=this.third();
             System.out.println("third 返回的结果是: " + third);
            if(third >50){
                this.first();
                System.out.println("third 返回的结果>50,继续运行 this.first();" );
            }else
                System.out.println("因返回的结果<50,不再执行下一步操作。");
            }else
                System.out.println("因返回的结果<50,不再执行下一步操作。");
        }else
            System.out.println("因返回的结果<50,不再执行下一步操作。");
    }
}
```

修改后的 InstallSoftware 类:

```java
package com.迪米特法则;
public class  InstallSoftware {
    public void  installWizard(Wizard wizard){
        wizard.installWizard();
    }
}
```

客户端不变,运行结果也相同。

在重构后,Wizard 类对外只公布了一个 public 方法 installWizard(),即使要修改 first 的返回值,影响的也只是 Wizard 本身,其他类不受影响,这显示了类的高内聚的特性。

注意:迪米特法则要求类"羞涩"一点,尽量不要对外公布太多的 public 方法和非静态的 public 变量,尽量内敛,多使用 private、protected 等访问权限。

1.7 单一职责原则

对一个类而言,应该仅有一个原因可以引起它的变化。如果你能想到多于一个的原因改变一个类,那么这个类就具有多个职责。应该把多余的职责分离出去,分别创建一些类来完成每一个职责。

在编程过程中经常能遇上单一职责原则(Single Responsibility Principle,SRP),比如,修改用户名和密码,可能的写法如下:

```
void change(String userName,String address);
```

如果只需要修改一个数据,上面的代码也要求必须将两个数据都传进去,这就违反了单一职责原则。原因很简单:方法职责不单一。正确的写法应该是:

```
void changeName(String userName);
void changeAddress(String address);
```

单一职责原则备受争议,争议之处就是对职责的定义,什么是类的职责,以及怎么划分类的职责。我们先举个例子来说明什么是单一职责原则。

1.7.1 应用实例:用户信息管理系统

用户信息管理系统涉及用户、机构、角色管理这些模块,基本上使用的是 RBAC 模型(基于角色的访问控制 Role-Based Access Control,RBAC)。

1. 违反单一职责原则的设计

用户信息管理系统包括管理用户、修改用户的信息、增加机构(一个人属于多个机构)、增加角色等。用户有这么多信息和行为要维护,一种思路是把这些信息和行为写到一个接口中,因为都是用户管理类。按此思路设计的系统结构如图 1-29 所示。

可以看出这个接口设计得有问题,用户的属性(Property)和用户的行为(Behavior)没有分开,这是一个严重的错误!

2. 遵守单一职责原则的设计

图 1-29 中的接口设计得一团糟,应该把用户的信息抽取成一个 BO(Business Object,业务对象),把行为抽取成一个 BL(Business Logic,业务逻辑)。

正确的设计应该是:接口 IUserBO 负责用户的属性,职责是收集和反馈用户的属性信息;IUserBL 负责用户的行为,完成用户信息的维护和变更。

以上把一个接口拆分成两个接口的设计,就是依据单一职责原则。那什么是单一职责原则呢?单一职责原则的定义是:应该有且仅有一个原因引起类或

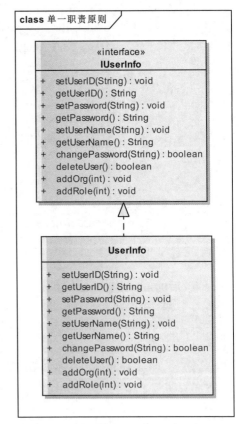

图 1-29 违反单一职责原则的系统结构

类中方法的变更。按照这个思路对类图进行修正，职责划分后的系统结构如图 1-30 所示。

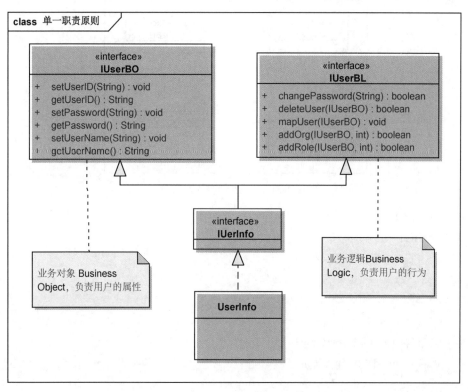

图 1-30　遵守单一职责原则的系统结构

我们现在是面向接口编程，所以在产生了这个 UserInfo 对象之后，可以把它当作 IUserBO 接口使用。当然，也可以当作 IUserBL 接口使用，这要取决于你在什么地方使用了。

示例代码如下：

```
........
IUserInfo userInfo = new UserInfo();
//我要赋值了，我就认为它是一个纯粹的 BO
IUserBO userBO = (IUserBO)userInfo;
userBO.setPassword("abc");
//我要执行动作了，我就认为是一个业务逻辑类
IUserBL userBL = (IUserBL)userInfo;
userBL.deleteUser();
........
```

要获得用户信息，就把 UserInfo 对象当作 IUserBO 的实现类，如代码第 3、4 行所示。如果希望维护用户的信息，就把它当作 IUserBL 的实现类，如代码第 6、7 行所示。

1.7.2　用户信息管理系统设计与 Java 实现

在实际使用中，我们更倾向于分别实现两个不同的类或接口：IUserBO 和 IUserBL，这时改进后的系统结构如图 1-31 所示。

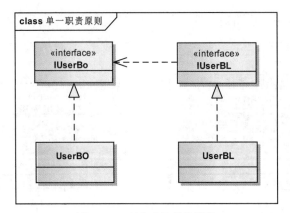

图 1-31 改进后的系统结构

改进后的代码如下。

接口 IUserBO：

```
package com.单一职责原则.改进;
public interface IUserBO {
    void setPassword(String string);
}
```

实现接口 IUserBO 的 UserBO 类：

```
package com.单一职责原则.改进;
public class UserBO implements IUserBO {
public void setPassword(String string){
        System.out.println("设置的密码是: " + string);
    }
}
```

接口 IUserBL：

```
package com.单一职责原则.改进;
public interface IUserBL {
    void deleteUser();
    void setPassword(IUserBO userBO, String string);
}
```

实现接口 IUserBL 的 UserBL 类：

```
package com.单一职责原则.改进;
public class UserBL implements IUserBL {
    public void deleteUser() {
        System.out.println("密码被删除！");
    }
    public void setPassword(IUserBO userBO,String string) {
        userBO.setPassword(string);
    }
}
```

测试程序如下：

```
package com.单一职责原则.改进;
public class Test {
    public static void main(String[] args) {
        IUserBL userBL = new UserBL();
        IUserBO userBO = new UserBO();
        userBL.setPassword(userBO,"abc");
        userBL.deleteUser();
    }
}
```

测试程序运行结果如图 1-32 所示。

图 1-32　遵守单一职责原则的用户信息管理系统 Java 实现运行结果

1.7.3　用户信息管理系统 Python 实现

```
import abc
class IUserBO:
    @abc.abstractmethod
    def setPassword(self,password):
        pass


class UserBO(IUserBO):
    def __init__(self) -> object:
        pass


    def setPassword(self,password):
        print("设置的密码是: " + password)

class IUserBL(metaclass=abc.ABCMeta):
    @abc.abstractmethod
    def deleteuser(self):
        pass
    # @abc.abstractmethod
    def setPassword(self,userbo,userbl):
        pass

class UserBL(IUserBL):
    def __init__(self,password) -> object:
        self.p = password
```

```python
        self.UserBO = UserBO()

    def showPassword(self):
        print("原来的的密码是: ", self.p)

    def deleteuser(self):
        print("密码被删除! ")
self.p = ''

    def setPassword(self,password):
        self.UserBO.setPassword(password)

if __name__ == '__main__':
    ubl = UserBL('abc')
    ubl.showPassword()
    ubl.setPassword('password')
    ubl.deleteuser()
    ubl.showPassword()
```

程序运行结果如图 1-33 所示。

图 1-33　遵守单一职责原则的用户信息管理系统 Python 实现运行结果

1.8　UML 简介

本节介绍 UML 中的六大关系（依赖、关联、聚合、组合、泛化、实现）及对应的 Java 代码。

1.8.1　依赖

实体之间的依赖（Dependency）关系表示在一个实体的规范发生变化后，可能影响依赖于它的其他实例。更具体地说，依赖可以转换为对不在实例作用域内的一个类或对象的任何类型的引用。它包括一个局部变量，通过调用方法获得对一个对象的引用，如图 1-34 所示，或者对一个类的静态方法的引用（同时不存在那个类的一个实例）。

依赖也可以用来表示包和包之间的关系。由于包中含有类，所以可以根据那些包中的各个类之间的关系，表示包和包之间的关系。

第 1 章 软件设计原则与 UML 简介

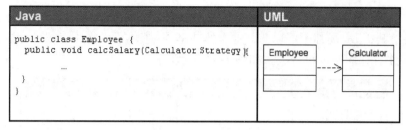

图 1-34 依赖关系

1.8.2 关联

实体之间的结构化关系表明对象是相互连接的。箭头是可选的，它用于指定导航能力。如果没有箭头，则暗示是一种双向的导航能力。在 Java 中，关联（Association）转换为一个实例作用域的变量，如图 1-35 所示。

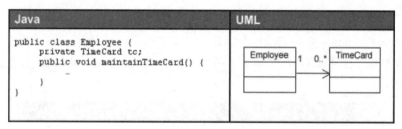

图 1-35 关联关系

可以为一个关联附加其他修饰符。多重性（Multiplicity）修饰符暗示着实例之间的关系。在示例代码中，Employee 可以有 0 个或更多的 TimeCard 对象。但是，每个 TimeCard 对象只从属于一个 Employee。

1.8.3 聚合

聚合（Aggregation）是关联的一种形式，代表两个类之间的整体或局部关系。聚合暗示整体在概念上位于比局部更高的级别，而关联暗示两个类在概念上位于相同的级别。聚合也转换为 Java 中的一个实例作用域的变量，如图 1-36 所示。

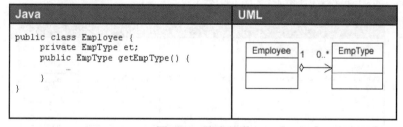

图 1-36 聚合关系

关联和聚合的区别纯粹是概念上的，而且严格反映在语义上。聚合还暗示不存在回路。换言之，聚合只能是一种单向关系。

1.8.4 组合

组合（Composition）是聚合的一种特殊形式，暗示局部在整体内部的生存期职责。组合是非共享的。虽然局部不一定要随整体的销毁而被销毁,但是整体要么负责保持局部的存活状态,要么负责将其销毁。局部不可以与其他整体共享,但整体可以将所有权转交给另一个对象,后者随即承担生存期职责。组合关系如图 1-37 所示。

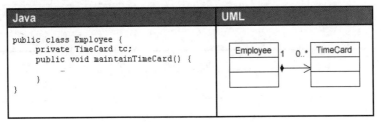

图 1-37　组合关系

Employee 和 TimeCard 的关系更适合表示成 "组合",而不是表示成 "关联"。

1.8.5　泛化

泛化(Generalization)关系表示一个泛化的元素和一个具体的元素之间的关系。泛化关系是用于对继承关系进行建模的 UML 元素。在 Java 中,直接用 extends 关键字表示这种关系,如图 1-38 所示。

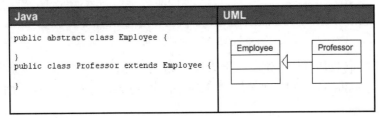

图 1-38　泛化关系

1.8.6　实现

实现(Realization)关系指定两个实体之间的一个合同(接口)。换言之,一个实体定义一个合同,而另一个实体保证履行该合同。在使用 Java 应用程序建模时,实现关系可以直接用 implements 关键字表示,如图 1-39 所示。

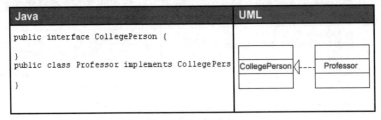

图 1-39　实现关系

1.9　设计模式简介

设计模式(Design Pattern)是一套多数人知晓的、被反复使用的、经过分类编目的代码设计经验的总结。使用设计模式是为了可重用代码,让代码更容易被他人理解,保证代码的可靠性。毫无疑问，设计模式于己于他人于系统是共赢的，设计模式使编写代码真正工程化，它是软件工程的基石，如同大厦的骨架。

设计模式四要素是模式名称、问题、解决方案、效果。

设计模式分为三大类型，本书只讲解其中最常用的 14 种设计模式。

第 2 章介绍 5 种创建型模式：简单工厂模式、工厂方法模式、抽象工厂模式、单例模式、多例模式。

第 3 章介绍 4 种结构型模式：适配器模式、默认适配器模式、装饰模式、门面模式。

第 4 章介绍 5 种行为型模式：策略模式、模板方法模式、命令模式、状态模式、观察者模式。

第 2 章 创建型模式

创建型模式（Creational Pattern）是对类的实例化过程的抽象化。

一些系统在创建对象时，需要动态地决定怎样创建对象，创建哪些对象，以及如何组合和表示这些对象。创建型模式描述了怎样构造和封装这些动态的决定。

创建型模式分为类的创建型模式和对象的创建型模式两种。

（1）类的创建型模式。类的创建型模式使用继承关系，把类的创建延迟到子类，从而封装了客户端将得到的具体类的信息，并且隐藏了这些类的实例是如何被创建和组合在一起的。

（2）对象的创建型模式。对象的创建型模式是把对象的创建过程动态地委派给另一个对象，从而动态地决定客户端将得到哪些具体类的实例，以及这些类的实例是如何被创建和组合在一起的。

本章将要介绍的创建型模式包括以下几种：简单工厂模式、工厂方法模式、抽象工厂模式、单例模式和多例模式。

工厂模式专门负责将大量有共同接口的类实例化。工厂模式可以动态决定将哪一个类实例化，不必事先知道每次要实例化哪一个类。工厂模式有以下几种形态。

（1）简单工厂模式（Simple Factory Pattern），又称静态工厂方法模式（Static Factory Method Pattern）。

（2）工厂方法模式（Factory Method Pattern），又称多态性工厂模式（Polymorphic Factory Pattern）或虚拟构造子模式（Virtual Constructor Pattern）。

（3）抽象工厂模式（Abstract Factory Pattern），又称工具箱（Kit 或 Toolkit）模式。

2.1 简单工厂模式

简单工厂模式（Simple Factory Pattern）是类的创建型模式，又被称为静态工厂方法模式（Static Factory Method Pattern）。简单工厂模式是由一个工厂对象决定创建哪一种产品类的实例。

简单工厂模式是工厂模式家族中最简单实用的模式，可以将其理解为不同工厂模式的一个特殊实现。

简单工厂模式将类实例化的操作与使用对象的操作分开，让使用者不需要知道具体参数就可以实例化出需要的"产品"类，从而避免了在客户端代码中的显式指定，实现了解耦。

简单工厂模式的缺点是当产品被修改时，工厂类也要做相应的修改。

简单工厂模式的适用场景有以下 3 种。

(1) 工厂类负责创建的对象比较少，由于创建的对象较少，不会造成工厂方法中的业务逻辑太过复杂

(2) 客户端只知道传入工厂类的参数，对如何创建对象并不关心

(3) 不确定会有多少个处理操作，例如，针对相同的数据，处理的逻辑可能会不同，可能以后还会增加新的操作

2.1.1 简单工厂模式的结构

简单工厂模式可根据传入的参数决定应创建哪个类的实例。简单工厂模式的一般结构如图 2-1 所示。

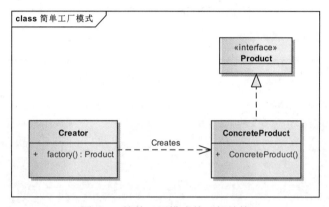

图 2-1 简单工厂模式的一般结构

从图 2-1 可以看出，简单工厂模式由工厂角色、抽象产品角色、产品角色 3 部分组成。

(1) 工厂（Creator）角色：是简单工厂模式的核心，但客户端调用它的工厂方法的时候，返回给客户端的是产品角色的一个类的对象实例。

(2) 抽象产品（Product）角色：定义了产品角色的共性，通常由一个 Java 接口或一个 Java 抽象类定义。如果具体产品之间没有共同的商业逻辑，就使用 Java 接口，如果有共同的商业逻辑，就使用 Java 抽象类。

(3) 产品（ConcreteProduct）角色：使用简单工厂模式创建的任何一个对象都是这个角色的一个类的对象实例。

简单工厂模式的角色之间是可以相互合并的，例如 3 个角色合并成 1 个，就好像单例模式，单例模式自身就是自己的工厂角色，但合并后的简单工厂模式并不完全等同于单例模式，单例模式中的构造方法是私有的。

简单工厂模式的缺点是当在产品角色中再增加一个类的时候，工厂角色必须发生相应的改变，这就导致该模式的扩展性不符合开闭原则。

2.1.2 应用系统登录 Java 实现

简单工厂模式在什么场景下使用呢？下面以登录功能为例说明。

假如应用系统需要支持多种登录方式，如口令认证、域认证（口令认证通常是去数据库中验证用户，而域认证则需要在指定的域中验证用户）。自然的做法是建立一个各种登录方式都适用的接口，如图 2-2 所示。

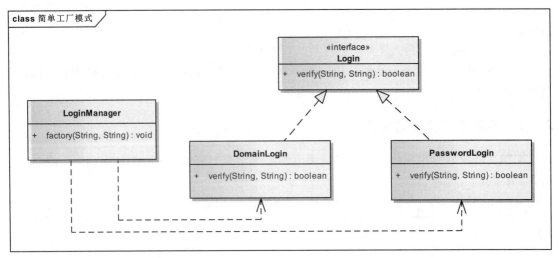

图 2-2　各种登录方式都适用的接口

抽象产品（Creator）角色：

```
package com.simpleFactory;
public interface Login {
    //登录验证
    public boolean verify(String name , String password);
}
```

产品（ConcreteProduct）角色：

```
package com.simpleFactory;
public class  DomainLogin implements Login {
    public boolean verify(String name, String password) {
        //业务逻辑
        System.out.println("登录类名为： DomainLogin");
        return true;
    }
}
```

产品（ConcreteProduct）角色：

```
package com.simpleFactory;
public class  PasswordLogin implements Login {
    public boolean verify(String name, String password) {
        //业务逻辑
        System.out.println("登录类名为： PasswordLogin");
        return true;
    }
}
```

我们还需要一个工厂类 LoginManager，根据调用者不同的要求，创建不同的登录对象并返回。如果碰到不合法的要求，会返回一个 Runtime 异常。

工厂（Creator）角色：

```java
package com.simpleFactory;
public class LoginManager {
    public static Login factory(String type){
        if(type.equals("password")){
            return new PasswordLogin();
        }else if(type.equals("passcode")){
            return new DomainLogin();
        }else{
            //这里抛出一个自定义异常会更恰当
            throw new RuntimeException("没有找到登录类型");
        }
    }
}
```

测试程序如下：

```java
package com.simpleFactory;
public class Client {
    public static void main(String[] args) {
//登录验证字符串"password"和"passcode"能正确运行，其他字符串将抛出异常
        loginMethod("password");
        loginMethod("passcode");
        loginMethod("abc");
    }
    public static void loginMethod(String loginType){
        String name = "name";
        String password = "password";
        Login login = LoginManager.factory(loginType);
        boolean bool = login.verify(name, password);
        if (bool) {
            //业务逻辑
            System.out.println("登录成功");
        }else {
            //业务逻辑
            System.out.println("失败");
        }
    }
}
```

程序运行结果如图 2-3 所示。

运行结果表明，程序能按要求正确运行，loginMethod("abc")也因"abc"属于其他字符串而抛出了异常。

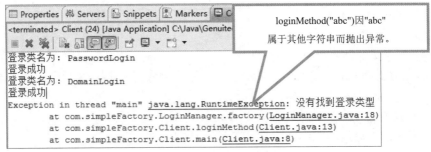

图 2-3 应用系统登录 Java 实现程序运行结果

2.1.3 简单工厂模式的优缺点

1. 简单工厂模式的优点

简单工厂模式的核心是工厂类。这个类含有必要的逻辑判断，可以决定在什么时候创建哪一个登录验证类的实例，因此，调用者可以免除直接创建对象的责任。简单工厂模式通过这种做法实现了对责任的分割，当系统引入新的登录方式的时候无须修改调用者。

2. 简单工厂模式的缺点

工厂类集中了所有的创建逻辑，当有复杂的多层次等级结构时，所有的业务逻辑都在工厂类中实现。当工厂类不能工作的时候，整个系统都会受到影响。

2.1.4 练习

1. 农场信息管理系统 Java 实现

有一个农场，专门向市场销售各类水果，在这个农场的信息管理系统中需要描述 3 种水果：葡萄（Grape）、草莓（Strawberry）、苹果（Apple）。

水果与其他植物不同，最终可以采摘食用，一个自然的做法是建立一个各种水果都适用的接口，将水果和农场里的其他植物区分开，农场信息管理系统的设计方案如图 2-4 所示。

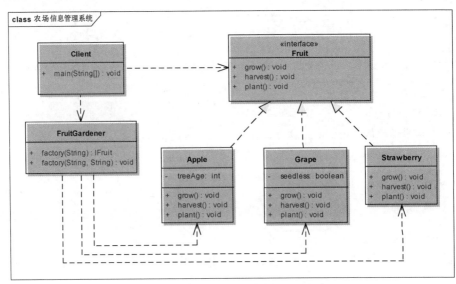

图 2-4 农场信息管理系统的设计方案

源代码如下。
抽象产品角色（Product）：

```java
package com.simpleFactory.fruit;
public interface Fruit {
    //生长
    void grow();
    //收获
    void harvest();
    //种植
    void plant();
}
```

产品角色（ConcreteProduct）1：

```java
package com.simpleFactory.fruit;
public class Apple implements Fruit {
    private int treeAge;
    @Override
    public void grow() {
        log("苹果正在生长中，请耐心等待...");
    }
    @Override
    public void harvest() {
        log("苹果已到收获季，请收获...");
    }
    @Override
    public void plant() {
        log("苹果可以种植，请悉心照料...");
    }
    private static void log(String msg) {
        System.out.println(msg);
    }
    public int getTreeAge() {
        return treeAge;
    }
    public void setTreeAge(int treeAge) {
        this.treeAge = treeAge;
    }
}
```

产品角色（ConcreteProduct）2：

```java
package com.simpleFactory.fruit;

public class Grape implements Fruit {
    booleanseedless ;
```

```java
    @Override
    public void grow() {
        Display("葡萄正在生长中,请等待......");
    }
    @Override
    public void harvest() {
        Display("葡萄已到收获季,请收获......");
    }
    @Override
    public void plant() {
        Display("葡萄可以种植,请悉心照料......");
    }
    private static void Display(String msg) {
        System.out.println(msg);
    }
    public boolean getSeedless() {
        return seedless;
    }
    public void setSeedless(boolean seedless) {
        this.seedless = seedless;
    }
}
```

产品角色(ConcreteProduct)3:

```java
package com.simpleFactory.fruit;
public class Strawberry implements Fruit {
    @Override
    public void grow() {
        Display("草莓正在生长中,请等待......");
    }
    @Override
    public void harvest() {
        Display("草莓已到收获季,请收获......");
    }
    @Override
    public void plant() {
        Display("草莓可以种植,请悉心照料......");
    }
    private static void Display(String msg) {
        System.out.println(msg);
    }
}
```

工厂角色(Creator):

```java
package com.simpleFactory.fruit;
```

```java
public class FruitGardener {
    public static Fruit factory(String fruit) throws BadFruitException{
        if(fruit.equalsIgnoreCase("apple"))
            return new Apple();
        else if(fruit.equalsIgnoreCase("strawberry"))
            return new Strawberry();
        else if(fruit.equalsIgnoreCase("grape"))
            return new Grape();
        else
            throw new BadFruitException("Bad fruit request");
    }
    public static void factory(String fruit,String farmwork) throws BadFruitException{
        if(farmwork.equalsIgnoreCase("grow")) {
            FruitGardener.factory(fruit).grow();
        }
        else if(farmwork.equalsIgnoreCase("harvest")) {
            FruitGardener.factory(fruit).harvest();
        }
        else if(farmwork.equalsIgnoreCase("plant")) {
            FruitGardener.factory(fruit).plant();
        }else if(farmwork.equalsIgnoreCase("strawberry")) {
            FruitGardener.factory(fruit).plant();
        }else if(farmwork.equalsIgnoreCase("grape")) {
            FruitGardener.factory(fruit).plant();
        }
        else
            throw new BadFruitException("Bad fruit request");
    }
}
```

输入异常处理：

```java
package com.simpleFactory.fruit;

public class BadFruitException extends Exception {
    public BadFruitException(String msg) {
        super(msg);
    }
}
```

客户端：

```java
package com.simpleFactory.fruit;
import java.util.Scanner;
public class Client {
```

```java
public static void main(String[] args) {
    Scanner scanner = new Scanner(System.in);

    while (true) {
        String msg1 = null;
        String msg2 = null;
        String msg3 = null;
        System.out.println("请输入水果和状态: ");
        msg1 = scanner.next();
        msg2 = scanner.next();
        try {
            FruitGardener.factory(msg1, msg2);
        } catch (BadFruitException e) {
            System.out.println("很抱歉,您的输入有误,请检查。");
        }
        System.out.println("是否继续?(Y/N)");
        msg3 = scanner.next().substring(0);
        if (msg3.equalsIgnoreCase("N"))
            System.exit(0);
    }
}
```

程序运行结果如图 2-5 所示。

图 2-5 简单工厂模式农场信息管理系统 Java 实现程序运行结果

2. 农场信息管理系统 Python 实现

```python
# 父类
import abc
class Fruit (object):
    @abc.abstractmethod
    def grow(self):
        pass
    @abc.abstractmethod
```

```python
        def harvest(self):
            pass
        @abc.abstractmethod
        def plant(self):
            pass
# Fruit 子类
class Apple (Fruit):
    __treeAge = 0    #类的私有属性
    def __init__(self):
        print("现在开始种苹果树!")
        self.__treeAge = 0   #对象的私有属性
        print("苹果树龄为: ", self.__treeAge)
    def grow(self):
        print("苹果正在生长中,请耐心等待...")

    def harvest(self):
        print("苹果已到收获季,请收获...\n")
    def plant(self):
        print("苹果可以种植,请悉心照料...")
    def getTreeAge(self):
        print("苹果树龄为: " , self.__treeAge)
        return self.__treeAge
    def setTreeAge(self,treeAge):
        self.__treeAge = treeAge


# Fruit 子类
class Grape (Fruit):
    def __init__(self):
        self.grapeAge = 0

    def grow(self):
        print("葡萄 正在生长中,请耐心等待...")
    def harvest(self):
        print("葡萄 已到收获季,请收获...\n")
    def plant(self):
        print("葡萄 可以种植,请悉心照料...")
'''
简单工厂模式:按需调用子类中的方法
用户可通过该类选择 Shape 的子类进行实例化
'''
class ShapeFactory(object):
    def create(self, shape):
        if shape == 'apple':
```

```python
            return Apple()
        elif shape == 'grape':
            return Grape()
        else:
            return None
```

测试程序如下:

```python
if __name__ == '__main__':
    fac = ShapeFactory()   # 实例化工厂类
    obj = fac.create('apple')   # 实例化 Shape 的 Circle 子类
    obj.grow()
    print("苹果树龄为: ", obj.__treeAge)   # 类和对象的私有属性都不能访问
    obj.setTreeAge(12)
    obj.getTreeAge()
    obj.harvest()

    fac = ShapeFactory()   # 实例化工厂类
    obj = fac.create('grape')   # 实例化 Shape 的 Circle 子类
    obj.grow()
    obj.harvest()
```

程序运行结果如图 2-6 所示。

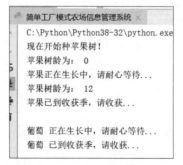

图 2-6　简单工厂模式农场信息管理系统 Python 实现程序运行结果

2.2　工厂方法模式

工厂方法模式（Factory Method Pattern）是类的创建型模式，又称为虚拟构造子模式（Virtual Constructor Pattern）或多态性工厂模式（Polymorphic Factory Pattern）。

工厂方法模式定义一个创建产品对象的工厂接口，与简单工厂模式的区别是将实际创建工作交给子类。

在工厂方法模式中，核心工厂类不再负责创建所有产品，而是将具体创建的工作交给子类，成为一个抽象工厂角色，仅负责给出具体工厂类必须实现的接口，而不涉及哪一个产品类应当被实例化这种细节。

工厂方法模式的适用场景有以下两种。
（1）客户端不知道它需要的对象的类。
（2）抽象工厂类通过其子类指定创建对象。

2.2.1　工厂方法模式的结构

工厂方法模式的结构如图 2-7 所示。

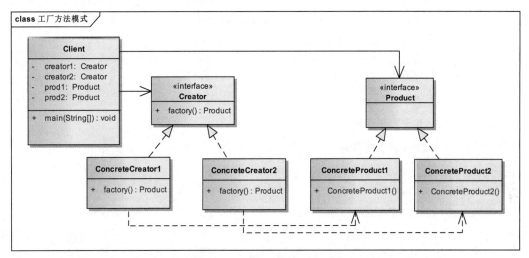

图 2-7　工厂方法模式的结构

从图中可以看出，这个使用了工厂方法模式的系统涉及以下角色。

（1）抽象工厂（Creator）角色：这个角色是工厂方法模式的核心，它与应用程序无关。在模式中，任何创建对象的工厂类必须实现这个接口。在图 2-7 中，这个角色由 Creator 接口扮演。在实际的系统中，这个角色也经常用抽象类实现。

（2）具体工厂（ConcreteCreator）角色：担任这个角色的是实现了抽象工厂接口的具体 Java 类。具体工厂角色含有与应用密切相关的逻辑，并且被应用程序调用以创建产品对象。在图 2-7 中给出了两个这样的角色，也就是具体 Java 类 ConcreteCreator1 和 ConcreteCreator2。

（3）抽象产品（Product）角色：这个角色是工厂方法模式创建的对象（即产品对象）的共同父类或共同拥有的接口。在图 2-7 中，这个角色为 Product 接口。

（4）具体产品（ConcreteProduct）角色：这个角色实现了抽象产品角色声明的接口。工厂方法模式创建的每一个对象都是某个具体产品角色的实例。在图 2-7 中，这个角色由具体类 ConcreteProduct1 和 ConcreteProduct2 扮演，它们都实现了 Product 接口。

工厂方法模式结构的示意性源代码如下。

抽象工厂（Creator）角色：

```
package com.factorymethod;
public interface Creator{
    public Product factory();
}
```

具体工厂（ConcreteCreator）角色 1：

```
package com.factorymethod;
```

```java
public class ConcreteCreator1 implements Creator{
    public Product factory(){
        return new ConcreteProduct1();
    }
}
```

具体工厂（ConcreteCreator）角色2：

```java
package com.factorymethod;
public class ConcreteCreator2 implements Creator{
    public Product factory(){
        return new ConcreteProduct2();
    }
}
```

抽象产品（Product）角色：

```java
package com.factorymethod;
public interface Product {
}
```

具体产品（ConcreteProduct）角色1：

```java
package com.factorymethod;
public class ConcreteProduct1 implements Product{
    public ConcreteProduct1(){
        System.out.println("ConcreteProduct1 is being created.");
    }
}
```

具体产品（ConcreteProduct）角色2：

```java
package com.factorymethod;
public class ConcreteProduct2 implements Product{
    public ConcreteProduct2(){
        System.out.println("ConcreteProduct2 is being created.");
    }
}
```

测试程序如下：

```java
package com.factorymethod;
public class Client {
    private static Creator creator1, creator2;
    private static Product prod1, prod2;
    public static void main(String[] args) {
        creator1 = new ConcreteCreator1();
        prod1 = creator1.factory();
        creator2 = new ConcreteCreator2();
        prod2 = creator2.factory();
```

　　　　}
　　}

测试程序运行结果如图 2-8 所示。

```
Problems  @ Javadoc  Declaration
<terminated> Client (25) [Java Application] C:\Ja
ConcreteProduct1 is being created.
ConcreteProduct2 is being created.
```

图 2-8　工厂方法模式程序运行结果

2.2.2　练习

1. 农场信息管理系统的设计与 Java 实现

1）系统需求分析

现在继续以农场的信息管理系统为例。在本章的"简单工厂模式"一节里，讨论了支持水果类作物的系统。在那个系统中，有一个全知全能的园丁角色 FriutGardener，这个角色控制所有作物的种植、生长和收获。现在这个农场的规模变大了，要求管理更加专业化。过去的全能人物没有了，每一种农作物都由专门的园丁管理，形成规模化和专业化生产。

2）系统设计

取代了过去的全能角色的是一个抽象的园丁角色，这个角色规定具体园丁角色需要实现的具体职能，而真正负责作物管理的是负责各种作物的具体园丁角色。

这一节仍然选择前面所讨论过的植物，包括葡萄（Grape）、草莓（Strawberry）以及苹果（Apple）。专业化的管理要求有专门的园丁负责专门的水果，比如苹果由"苹果园丁"负责，草莓由"草莓园丁"负责，而葡萄由"葡萄园丁"负责。这些"苹果园丁""草莓园丁""葡萄园丁"都是实现了抽象的"水果园丁"接口的具体工厂类，而"水果园丁"则扮演抽象工厂角色。

这样一来，农场信息管理系统的结构如图 2-9 所示。

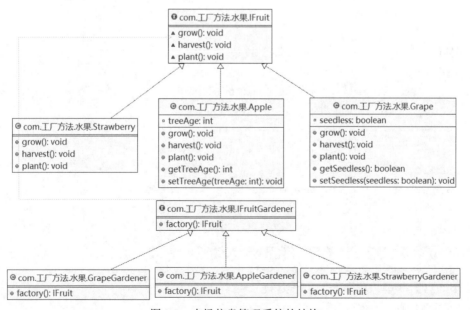

图 2-9　农场信息管理系统的结构

源代码如下：

```java
package com.工厂方法.水果;
public interface IFruit {
    void grow();
    void harvest();
    void plant();
}
//具体产品（Concrete Product）角色Apple
package com.工厂方法.水果;
public class Apple implements IFruit{
    private int treeAge;
    public void grow() {
     System.out.println("Apple is growing...");
    }
    public void harvest() {
     System.out.println("Apple has been harvested.");
    }

    public void plant() {
        System.out.println("Apple has been planted.");
    }
    public int getTreeAge() {
        return treeAge;
    }
    public void setTreeAge(int treeAge) {
        this.treeAge = treeAge;
    }
}
//具体产品（Concrete Product）角色Grape
package com.工厂方法.水果;
public class Grape implements IFruit {
    public void grow() {
     System.out.println("Grape is growing...");
    }
    public void harvest() {
     System.out.println("Grape has been harvested.");
    }
    public void plant() {
        System.out.println("Grape has been planted.");
    }
    public boolean getSeedless() {
        return seedless;
```

```java
    public void setSeedless(boolean seedless) {
        this.seedless = seedless;
    }
    private boolean seedless;
}
//具体产品（Concrete Product）角色Strawberry
package com.工厂方法.水果;
public class Strawberry implements IFruit {
    public void grow() {
        System.out.println("Strawberry is growing...");
    }
    public void harvest() {
        System.out.println("Strawberry has been harvested.");
    }
    public void plant() {
        System.out.println("Strawberry has been planted.");
    }
}
//抽象工厂（Creator）角色FruitGardener
package com.工厂方法.水果;
public interface IFruitGardener {
    public IFruit factory();
}
//具体工厂（ConcreteCreator）角色AppleGardener
package com.工厂方法.水果;
public class AppleGardener implements IFruitGardener {
    public IFruit factory() {
        return new Apple();
    }
}
//具体工厂（ConcreteCreator）角色GrapeGardener
package com.工厂方法.水果;
public class GrapeGardener implements IFruitGardener {
    public IFruit factory() {
        return new Grape();
    }
}
//具体工厂（ConcreteCreator）角色StrawberryGardener
package com.工厂方法.水果;
public class StrawberryGardener implements IFruitGardener {
    public IFruit factory() {
        return new Strawberry();
```

        }
}
测试程序如下：
```java
package com.工厂方法.水果;
import java.util.Scanner;
public class Client {
    public static void main(String[] args) {
        Scanner scanner = new Scanner(System.in);
        while (true) {
            String fruitname = null;
            System.out.print("请输入水果名称： ");
            Scanner in=new Scanner(System.in);
            String a=in.next();
            IFruit iFruit=null;
            IFruitGardener iFruitGardener;
            if(a.equalsIgnoreCase("apple")){
                iFruitGardener=new AppleGardener();
                iFruit=iFruitGardener.factory();
                Apple apple =(Apple) iFruit;
                apple.setTreeAge(5);
                System.out.println("苹果树的年龄为:"+apple.getTreeAge()+"年");
            }
            else if(a.equalsIgnoreCase("grape")){
                iFruitGardener=new GrapeGardener();
                iFruit=iFruitGardener.factory();
                Grape grape=(Grape)iFruit;
                grape.setSeedless(true);
      System.out.println(grape.getSeedless()==true?"葡萄无籽":"葡萄有籽");
            }
            else if(a.equalsIgnoreCase("strawberry")){
                iFruitGardener=new StrawberryGardener();
                iFruit=iFruitGardener.factory();
            }
            iFruit.grow();
            iFruit.harvest();
            iFruit.plant();
            System.out.print("是否继续？（Y/N）");
            fruitname = scanner.next().substring(0);
            if (fruitname.equalsIgnoreCase("N")){
                System.out.println("程序退出，再见！");
                System.exit(0);
            }
```

```
            }
        }
}
```

程序运行结果如图2-10所示。

```
<terminated> Client (8) [Java Application]
请输入水果名称:apple
苹果树的年龄为:5年
Apple is growing...
Apple has been harvested.
Apple has been planted.
是否继续?(Y/N) y
请输入水果名称:grape
葡萄无籽
Grape is growing...
Grape has been harvested.
Grape has been planted.
是否继续?(Y/N) y
请输入水果名称:strawberry
Strawberry is growing...
Strawberry has been harvested.
Strawberry has been planted.
是否继续?(Y/N) n
程序退出,再见!
```

图2-10 工厂方法模式农场信息管理系统Java实现程序运行结果

2. 农场信息管理系统Python实现

```python
# 接口产品基类
# 父类
class Fruit (object):

    def grow(self):
        pass
    def harvest(self):
        pass
    def plant(self):
        pass


# Fruit子类
class Apple (Fruit):
    __treeAge = 0    #类的私有属性
    def __init__(self):
        print("现在开始种苹果树!")
        self.__treeAge = 0    #对象的私有属性
        print("苹果树龄为: ", self.__treeAge)
    def grow(self):
        print("苹果正在生长中,请耐心等待...")
    
    def harvest(self):
        print("苹果已到收获季,请收获...\n")
    def plant(self):
```

```python
        print("苹果可以种植,请悉心照料...")
    def getTreeAge(self):
        print("苹果树龄为: " , self.__treeAge)
        return self.__treeAge
    def setTreeAge(self,treeAge):
        self.__treeAge = treeAge

# Fruit 子类
class Grape (Fruit):
    def __init__(self):
        self.grapeAge = 0

    def grow(self):
        print("葡萄正在生长中,请耐心等待...")
    def harvest(self):
        print("葡萄已到收获季,请收获...\n")
    def plant(self):
        print("葡萄可以种植,请悉心照料...")

# 接口工厂基类
class IFactory(object):

    def create(self):
        # 把要创建的工厂对象装配进来
        raise NotImplementedError
class AppleFactory(IFactory):
    def create(self):
        return Apple()

class GrapeFactory(IFactory):
    def create(self):
        return Grape()

if __name__ == "__main__":
    apple = AppleFactory()
    obj = apple.create()
    obj.plant()
    obj.grow()
    obj.setTreeAge(12)
    obj.getTreeAge()
    obj.harvest()

    grape = GrapeFactory()
```

```
    obj = grape.create()
    obj.plant()
    obj.grow()
    obj.harvest()
```

程序运行结果如图 2-11 所示。

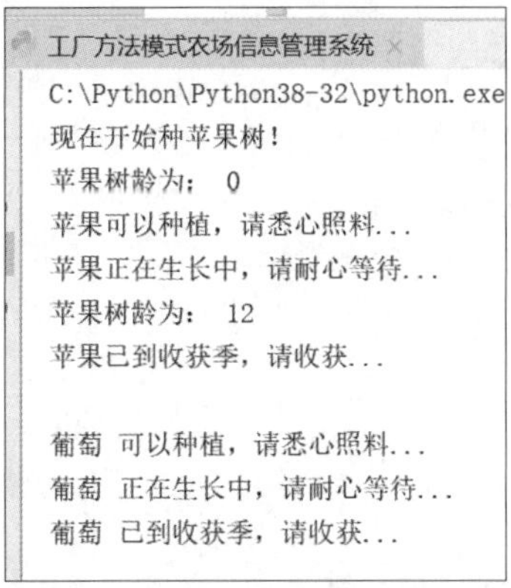

图 2-11 使用工厂方法模式的农场信息管理系统 Python 实现程序运行结果

3. 计算图形面积 Python 实现

```
import math
# 定义 4 个图形类,并且每一个图形都有一个可以计算面积的方法
class Circle:
    def Area(self, radius):
        return math.pow(radius, 2) * math.pi
class Rectangle:
    def Area(self, longth, width):
        return 2 * longth * width
class Triangle:
    def Area(self, baselong, height):
        return baselong * height / 2
class Ellipse:
    def Area(self, long_a, short_b):
        return long_a * short_b * math.pi
# 定义创建对象的工厂接口,因为 Python 中没有接口的概念,所以,这里通过类的继承实现
class IFactory:
    # 模拟接口
    def create_shape(self):
        # 定义接口的方法,只提供方法的声明,不提供方法的具体实现
        pass
class CircleFactory(IFactory):
# 模拟类型实现某一个接口,实际上是类的继承
    def create_shape(self,name):      # 重写接口中的方法
        if(name) == 'Circle':
            return Circle()
```

```python
class RectangleFactory(IFactory):
    def create_shape(self,name):        # 重写接口中的方法
        if(name) == 'Rectangle':
            return Rectangle()
class TriangleFactory(IFactory):
# 模拟类型实现某一个接口,实际上是类的继承
    def create_shape(self, name):       # 重写接口中的方法
        if(name) == 'Triangle':
            return Triangle()
class EllipseFactory(IFactory):
# 模拟类型实现某一个接口,实际上是类的继承
    def create_shape(self,name):        # 重写接口中的方法
        if(name) == 'Ellipse':
            return Ellipse()
if __name__ == '__main__':
    factory1 = CircleFactory()
    factory2 = RectangleFactory()
    factory3 = TriangleFactory()
    factory4 = EllipseFactory()

    circle = factory1.create_shape('Circle')
    circle_area = circle.Area(2)
    printf('这是一个圆,它的面积是:{circle_area}')

    rectangle = factory2.create_shape('Rectangle')
    rectangle_area = rectangle.Area(2, 3)
    printf('这是一个长方形,它的面积是:{rectangle_area}')

    triangle = factory3.create_shape('Triangle')
    triangle_area = triangle.Area(2, 3)
    printf('这是一个三角形,它的面积是:{triangle_area}')

    ellipse = factory4.create_shape('Ellipse')
    ellipse_area = ellipse.Area(3, 2)
    printf('这是一个椭圆,它的面积是:{ellipse_area}')
```

程序运行结果如图 2-12 所示。

```
工厂方法模式 求图形面积
C:\Python\Python38-32\python.exe "E:/教学资料/
这是一个圆,它的面积是:12.566370614359172
这是一个长方形,它的面积是:12
这是一个三角形,它的面积是:3.0
这是一个椭圆,它的面积是:18.84955592153876
```

图 2-12　计算图形面积 Python 实现程序运行结果

2.3 抽象工厂模式

抽象工厂模式（Abstract Factory Pattern）用于创建分属于不同操作系统如 UNIX、Windows 的构件，它可以向客户端提供一个接口，使客户端在不必指定产品具体类型的情况下，创建多个产品族中与自身匹配的产品对象。这就是抽象工厂模式的用意。

每种模式都是针对一定问题的解决方案。抽象工厂模式针对的问题是设计多产品等级结构的系统。

抽象工厂模式的定义：为创建一组相关的或相互依赖的对象提供一个接口，而且无须指定它们的具体类。

抽象工厂模式是由一个超级工厂创建其他工厂。该超级工厂又被称为其他工厂的工厂。这种设计模式属于创建型模式，它提供了一种创建对象的最佳方式。

意图：提供一个创建一系列相关的或相互依赖的对象的接口，而无须指定它们的具体类。

用途：主要解决选择接口的问题。

何时使用：在系统的产品有多个产品族，而系统只消费其中某一族的产品的时候。

如何解决：在一个产品族里定义多个产品。

关键代码：在一个工厂里聚合多个同类产品。

在抽象工厂模式中，接口负责创建一个相关对象的工厂，不需要显示指定它们的类。每个生成的工厂都能按照工厂方法模式提供对象。

在学习抽象工厂具体的实例之前，要先明白两个重要的概念：产品族和产品等级结构。

产品族是指位于不同产品等级结构中功能相关联的产品组成的家族。抽象工厂模式提供的一系列产品就组成一个产品族。如 Windows 操作系统和 Linux 操作系统就是两个不同的产品族。

工厂方法提供的一系列产品就是一个等级结构。例如，Windows 操作系统和 Linux 操作系统中的"按钮""文本框"分别为一个产品等级结构。

工厂方法模式和抽象工厂模式的区别：如果工厂的产品全部属于同一个产品族，则属于工厂方法模式；如果工厂的产品可来自多个产品族，则属于抽象工厂模式。

一般情况下，产品族和产品等级结构的关系如图 2-13 所示。

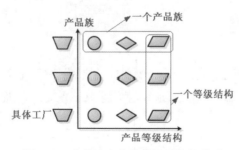

图 2-13　产品族和产品等级结构的关系

2.3.1 抽象工厂模式的起源

抽象工厂模式起源于 UNIX 操作系统和 Windows 操作系统，它们的构件如命令按键（Button）、文字框（Text）在不同的操作系统中有不同的本地实现，它们的细节也必然有所不同。

现在，有两个产品的等级结构，分别是 Button 等级结构和 Text 等级结构。同时有两个产品族，也就是 UNIX 产品族和 Windows 产品族，如图 2-14 所示。

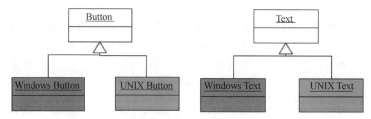

图 2-14　UNIX 产品族和 Windows 产品族

UNIX 产品族由 UNIX Button 和 UNIX Text 产品构成。而 Windows 产品族由 Windows Button 和 Windows Text 产品构成，如图 2-15 所示。

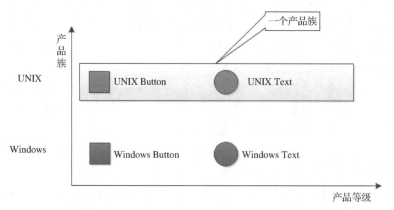

图 2-15　UNIX 和 Windows 产品族和产品等级结构

2.3.2　抽象工厂模式的结构

使用抽象工厂模式设计的系统结构如图 2-16 所示。

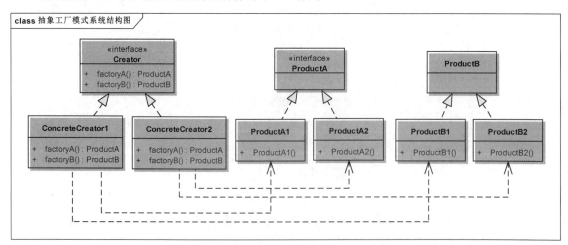

图 2-16　使用抽象工厂模式设计的系统结构

从图 2-16 可以看出，抽象工厂模式涉及以下角色。

（1）抽象工厂（AbstractFactory）角色：这个角色是抽象工厂模式的核心，它与应用系统的商业逻辑无关,通常使用 Java 接口或抽象 Java 类实现,所有的具体工厂类必须实现这个 Java 接口或继承这个抽象 Java 类。

(2)具体工厂类(ConcreteFactory)角色:这个角色直接在客户端的调用下创建产品的实例。这个角色含有选择合适的产品对象的逻辑,而这个逻辑与应用系统的商业逻辑紧密相关。通常使用具体 Java 类实现这个角色。

(3)抽象产品(AbstractProduct)角色:担任这个角色的类是抽象工厂模式创建的对象的父类,或它们共同拥有的接口。通常使用 Java 接口或抽象 Java 类实现这个角色。

(4)具体产品(ConcreteProduct)角色:抽象工厂模式创建的任何产品对象都是某一个具体产品类的实例,这是客户端最终需要的东西,其内部允满了应用系统的商业逻辑。通常使用具体 Java 类实现这个角色。

下面分析使用抽象工厂模式的 UNIX 和 Windows 产品族和产品等级结构的解决方案。

系统对产品对象的创建需求由一个工程的等级结构来满足,其中有两个具体工程角色,即 UNIXFactory 和 WindowsFactory。

UNIXFactory 对象负责创建 UNIX 产品族中的产品,而 WindowsFactory 对象负责创建 Windows 产品族中的产品,这就是抽象工厂模式的应用。

抽象工厂模式的解决方案如图 2-17 所示。

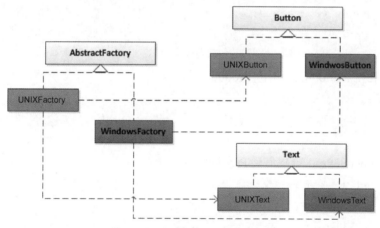

图 2-17 抽象工厂模式的解决方案

显然,一个系统只能在某一个操作系统的视窗环境下运行,不能同时在不同的操作系统中运行。所以,系统实际上只能消费属于同一个产品族的产品。

在现代的应用中,抽象工厂模式的使用范围已经扩大了,不再要求系统只能消费某一个产品族了。因此,可以不必理会前面提到的原始用意。

抽象工厂模式的示意性源代码如下。

抽象工厂(AbstractFactory)角色 Creator:

```
package com.abstractfactory;
public interface  Creator {
    public ProductA factoryA();
    public ProductB factoryB();
}
```

具体工厂类（ConcreteFactory）角色 ConcreteCreator1：

```java
package com.abstractfactory;
public class ConcreteCreator1 implements Creator {
public ProductA factoryA() {
   return new ProductA1();
 }
public ProductB factoryB() {
   return new ProductB1();
 }
}
```

具体工厂类（ConcreteFactory）角色 ConcreteCreator2：

```java
package com.abstractfactory;
public class ConcreteCreator2 implements Creator {
   public ProductA factoryA() {
      return new ProductA2();
   }
   public ProductB factoryB() {
      return new ProductB2();
   }
}
```

抽象产品（AbstractProduct）角色 ProductA：

```java
package com.abstractfactory;
public interface  ProductA {
}
```

具体产品（ConcreteProduct）角色 ProductA1：

```java
package com.abstractfactory;
public class  ProductA1 implements ProductA {
   public ProductA1() {
   }
}
```

具体产品（ConcreteProduct）角色 ProductA2：

```java
package com.abstractfactory;
public class  ProductA2 implements ProductA {
   public ProductA2() {
   }
}
```

抽象产品（AbstractProduct）角色 ProductB：

```java
package com.abstractfactory;
public interface  ProductB {
}
```

具体产品（ConcreteProduct）角色 ProductB1：

```
package com.abstractfactory;
public class ProductB1 implements ProductB {
   public ProductB1() {
   }
}
```

具体产品（ConcreteProduct）角色 ProductB2：

```
package com.abstractfactory;
public class ProductB2 implements ProductB {
   public ProductB2() {
   }
}
```

2.3.3 抽象工厂模式的优缺点

1. 抽象工厂模式的优点

（1）分离接口和实现。客户端使用抽象工厂创建需要的对象，而客户端根本就不知道具体的实现是谁，客户端只是面向产品的接口编程而已。也就是说，客户端从具体的产品实现中解耦。

（2）使切换产品族变得容易。因为一个具体的工厂实现代表的是一个产品族，比如上一小节例子中从 UNIX 系列到 Windows 系列只需要切换一下具体工厂。

2. 抽象工厂模式的缺点

不太容易扩展新的产品。如果需要给整个产品族添加一个新的产品，就需要修改抽象工厂，这样会导致修改所有的工厂实现类。

2.3.4 练习

1. 农场信息管理系统 Java 实现

1）农场信息管理系统需求分析

聪明的农场主总是让自己的庄园越来越有价值，"农场"在经历了简单工厂模式和工厂方法模式后，不断地扩大生产。如今，农场再次面临新的大发展，一项重要的工作就是引进塑料大棚技术，在大棚里种植热带（Tropical）和亚热带（Northern）的水果和蔬菜，以满足市场需求，获取更大的利益。

经过分析，在农场信息管理系统中，产品族及产品等级结构如图2-18所示。

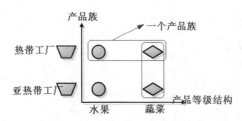

图 2-18　农场信息管理系统中的产品族及产品等级结构

2）系统设计

经过分析,所谓的各个园丁其实就是工厂角色,而蔬菜和水果则是产品角色。将抽象工厂模式用于农场中,系统的结构如图2-19所示。

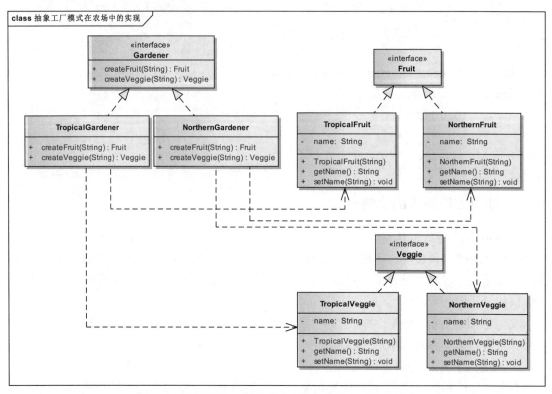

图2-19 使用抽象工厂模式的农场信息管理系统的结构

3）源代码

请见本书配套电子资源。

2. 农场信息管理系统 Python 实现

```python
from abc import abstractmethod, ABCMeta
class I水果(metaclass=ABCMeta):
    @abstractmethod
    def show(self):
        pass
class I蔬菜(metaclass=ABCMeta):
    @abstractmethod
    def show(self):
        pass
class 北方水果(I水果):
    def show(self):
        print("This is 北方水果")

class 南方水果(I水果):
    def show(self):
```

```python
        print("This is 南方水果")
class 北方蔬菜(I蔬菜):
    def show(self):
        print("This is 北方蔬菜")
class 南方蔬菜(I蔬菜):
    def show(self):
        print("This is 南方蔬菜")
class IFactory(metaclass=ABCMeta):
    @abstractmethod
    def create水果():
        pass
    @abstractmethod
    def create蔬菜():
        pass
class 南方Factory(IFactory):
    def create水果(self):
        return 南方水果()
    def create蔬菜(self):
        return 南方蔬菜()
class 北方Factory(IFactory):
    def create水果(self):
        return 北方水果()
    def create蔬菜(self):
        return 北方蔬菜()
class Client(object):
    def main(self):
        factory = 北方Factory()
        factory.create蔬菜().show()
        factory.create水果().show()
        factory = 南方Factory()
        factory.create蔬菜().show()
        factory.create水果().show()
if __name__ == "__main__":
    Client().main()
```

程序运行结果如图 2-20 所示。

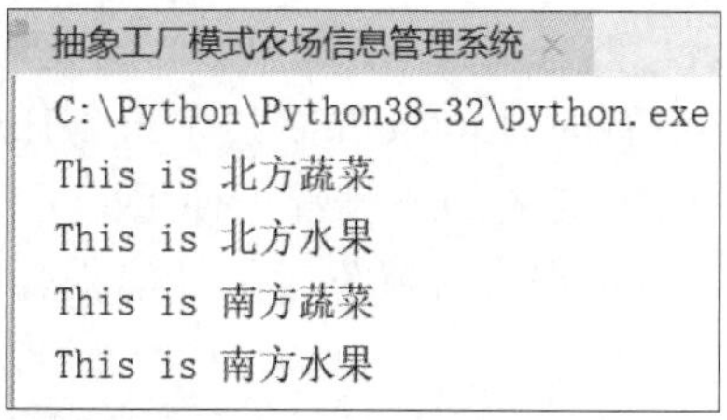

```
抽象工厂模式农场信息管理系统
C:\Python\Python38-32\python.exe
This is 北方蔬菜
This is 北方水果
This is 南方蔬菜
This is 南方水果
```

图 2-20 使用抽象工厂模式的农场信息管理系统 Python 实现程序运行结果

2.4 单例模式

作为对象的创建型模式,单例模式(Singleton Pattern)确保一个类只有一个实例,并且自行实例化并向整个系统提供这个实例,这个类称为单例类。

单例模式只有在有真正的"单一实例"的需求时才能使用。

应用实例:回收站

Windows 操作系统都有一个回收站,回收站自行提供自己的实例,在整个系统中,该回收站只能有一个实例,整个系统都使用这个唯一的实例。因此,回收站是单例模式的应用。

2.4.1 单例模式的结构

单例模式的特点如下。
(1)单例类只能有一个实例。
(2)单例类必须自己创建自己的唯一实例。
(3)单例类必须给其他对象提供这一实例。
单例模式的简略设计如图 2-21 所示。

2.4.2 单例模式常见的应用场景

(1)Windows 的任务管理器就是很典型的单例模式。
(2)Windows 的回收站。在整个系统运行过程中,回收站一直维护着仅有的一个实例。

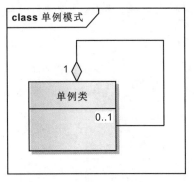

图 2-21 单例模式简略设计

(3)网站的计数器一般也采用单例模式实现,否则难以同步。
(4)应用程序的日志应用,一般都可以使用单例模式实现。这是由于共享的日志文件一直处于打开状态,因此只能由一个实例操作,否则不易追加内容。
(5)读取 Web 应用的配置对象,一般也采用单例模式。这是由于配置文件是共享的资源。
(6)数据库连接池的设计一般也采用单例模式,因为数据库连接是一种数据库资源。在数据库软件系统中使用数据库连接池,主要是为了降低在打开或关闭数据库连接时引起的效率损耗,这种效率上的损耗是非常昂贵的。使用单例模式维护数据库连接池可以大大降低这种损耗。
(7)设计多线程的线程池一般也采用单例模式,这是由于线程池要方便对池中的线程进行控制。
(8)操作系统的文件系统,也是单例模式实现的具体例子,一个操作系统只能有一个文件系统。
(9).Net 的 HttpApplication。熟悉 ASP.NET(IIS)的整个请求生命周期的人应该知道 HttpApplication 也是单例模式,所有的 HttpModule 共享一个 HttpApplication 实例。

总结以上,不难看出单例模式一般在以下情况下使用。
(1)在资源共享的情况下,避免在资源操作时的性能损耗。如上述中的日志文件,应用配置。
(2)在控制资源的情况下,方便资源之间的通信,如线程池等。

2.4.3 单例模式的类型

单例模式分为 3 种类型：饿汉式、懒汉式和登记式。下面分别讨论饿汉式和懒汉式。

1. 饿汉式单例模式

饿汉式是指在类装载时构建并初始化全局的单例实例，优点是速度快，在不调用函数时也能创建实例。饿汉式单例类被加载时，静态变量 instance 会被初始化，此时类的私有构造函数会被调用。

单例实例考虑线程安全因素后示意性程序如下：

```java
package com.singleton;
public class EagerSingleton {
    //该类只能有一个实例
    private EagerSingleton (){}   //私有无参构造方法
    //考虑线程安全,创建 ts1
    private static EagerSingleton ts1=null;
    public static EagerSingleton getTest(){
        if(ts1==null){
            ts1= new EagerSingleton ();
        }
        return ts1;
    }
}
```

测试程序如下：

```java
package com.singleton;
public class EagerSingletonTest {
    //@param args
    public static void main(String[] args) {
        EagerSingleton s=EagerSingleton.getTest();
        System.out.println("创建一个实例 s 的内存地址为: \t" + s);
        EagerSingleton s1=EagerSingleton.getTest();
        System.out.println("创建又一个实例 s1 的内存地址为: \t" + s1);
        if(s==s1){
            System.out.println("创建的是同一个实例");
        }else if(s!=s1){
            System.out.println("创建的不是同一个实例");
        }else{
            System.out.println("application error");
        }
    }
}
```

程序运行结果如图 2-22 所示。

图 2-22　饿汉式单例模式程序运行结果

在上面的例子中，在这个类被加载时，静态变量 instance 会被初始化，此时类的私有构造函数会被调用，单例类的唯一实例就被创建出来了。

饿汉式其实是一种比较形象的称谓。既然"饿"，那么在创建对象实例的时候就比较着急。饿了嘛，于是在装载类的时候就能创建对象实例。

```
private staticfinal EagerSingleton instance= new EagerSingleton();
```

饿汉式是典型的空间换时间，在类装载的时候就会创建类的实例，不管用不用，先创建出来，然后在每次调用的时候，就不需要再判断，节省了运行时间。

2. 懒汉式单例模式

懒汉式是指在第一次使用时构建全局的单例实例，延迟初始化。特点是速度慢，在调用时才创建单例实例。

1）单例模式懒汉式双重校验锁 Java 实现

```java
//单例模式懒汉式双重校验锁
package com.单例模式;
public class  LazySingleton {
    //懒汉式变种，属于懒汉式中最好的写法，保证了延迟加载和线程安全
    private static LazySingleton instance=null;
    private LazySingleton() {};
    public static LazySingleton getInstance(){
        if (instance == null) {
        synchronized (LazySingleton.class ) {
          if (instance == null) {
              instance = new LazySingleton();
            }
          }
        }
       return instance;
    }
}
```

上面这个懒汉式单例模式的实现对静态工厂方法进行了同步，以处理多线程环境。

懒汉式其实是一种比较形象的称谓。既然"懒"，那么就不着急创建对象实例，等到马上要使用对象实例的时候再创建。懒人总是在推脱不开的时候才会真正执行命令，因此在装载对象的时候不创建对象实例。

```
private static LazySingleton instance = null;
```

懒汉式是典型的时间换空间。在每次获取实例时都会对是否需要创建实例进行判断，浪

费判断的时间。当然，如果一直没有人使用的话，就不会创建实例，节约内存空间。为了保证懒汉式实现的线程安全，怎样以更好的方式实现呢？网上有很多相关解决方案，读者可自行查找。

2）懒汉式单例模式枚举 Java 实现

按照《Effective Java 中文版（第 2 版）》中的说法：单元素的枚举类型已经成为实现 Singleton 的最佳方法。用枚举实现单例模式非常简单，只需要编写一个包含单个元素的枚举类型。

```
package com.单例模式.枚举实现;
public class LazySingleton_enum {
    //构造方法私有化
    private LazySingleton_enum(){
    }
/**
 * 返回实例
 * @return
 */
public static LazySingleton_enum getInstance() {
return SingletonEnum.INSTANCE.getInstance();
    }
//使用枚举方法实现单例模式
privateenum SingletonEnum {
INSTANCE;
private LazySingleton_enum instance;

//JVM 保证这个方法只调用一次
    SingletonEnum() {
instance = new LazySingleton_enum();
    }

public LazySingleton_enum getInstance() {
    this.singletonOperation();
returninstance;
    }
//下面实现需要的方法
public void  singletonOperation(){
    int x = 3+2*9;
    System.out.print("\t3+2*9 = " + x);
    }
  }
}
```

测试程序如下：

```
package com.singleton;
public class LazySingleton_enumTest {
    public static void main(String[] args) {
```

```
        Singleton.getInstance();
    }
}
```

程序运行结果如图 2-23 所示。

图 2-23 枚举实现程序运行结果

使用枚举来实现单例模式实例的控制会更加简洁，而且枚举无偿地提供了序列化机制，并由 JVM 从根本上提供保障，防止多次实例化，是更简洁、高效、安全的实现单例模式的方式。

3）单例模式内部类 Java 实现

当一个单例类的初始化开销很大，且希望用户在实际需要的时候才创建单例类时，就会考虑使用懒汉式延迟初始化以提高程序的启动速度。另外在多线程的环境下，如果不同步 getInstance()方法，就会出现线程安全的问题，如果同步整个方法，那么 getInstance()就完全变成串行，串行效率会降低至原来的十分之一甚至百分之一。

继 enum 之后的最佳方法是使用静态内部类。在加载 singleton 时并不加载它的内部类 SingletonInner，而是在调用 getInstance()时，继而调用 SingletonInner 的情况下才加载 SingletonInner，从而调用 singleton 的构造函数，实例化 singleton，从而在不需要同步的情况下，达到延迟初始化的效果。

代码如下：

```
package com.singleton.进阶;
public class Singleton {
    private Singleton(){
    }
    //静态内部类
    static class SingletonInner{
        static Singleton instance = new Singleton();
    }
    public static Singleton getInstance(){
        return SingletonInner.instance;
    }

}
```

4）线程安全的单例模式 Python 实现

```
import threading
def synchronized(func):
    func.__lock__ = threading.Lock()
    def synced_func(*args, **kws):
        with func.__lock__:
            return func(*args, **kws)
    return synced_func
def Singleton(cls):
    instances = {}
    @synchronized
```

```python
    def get_instance(*args, **kw):
        if cls not in instances:
            instances[cls] = cls(*args, **kw)
        return instances[cls]
    return get_instance
def worker():
    single_test = test()
    print("id----> %s" % id(single_test))
@Singleton
class test():
    a = 1
if __name__ == "__main__":
    task_list = []
    for one in range(30):
        t = threading.Thread(target=worker)
        task_list.append(t)
    for one in task_list:
        one.start()
    for one in task_list:
        one.join()
```

程序运行结果如图 2-24 所示。

图 2-24　线程安全的单例模式 Python 实现程序运行结果

2.4.4　练习

如果 MySQL 数据库没有启动，可以在 Windows 10 系统中单击左下角的"开始"按钮，在"开始"菜单中选择"Windows PowerShell"（管理员）命令，输入"compmgmt.msc"后按回车键，在"计算机管理"窗口中选择"服务"选项，找到 MySQL 并启动。

1．连接 MySQL 数据库 Java 实现

连接 MySQL 数据库的源代码如下。

配置文件 jdbc.properties 的位置如图 2-25 所示。

```
class Name=com.mysql.jdbc.Driver
url=jdbc:mysql://localhost/mydb
username=root
password=123456
```

单例程序 GenericConnection：

```java
package com.singleton.单例程序连接MySQL;
import java.io.*;
import java.sql.*;
import java.util.Properties;
public class GenericConnection {
    private static Connection con;
    //单例获得连接的方法
    public static Connection getCon() {
        if(con == null) {
            String class Name, url, uid, password;
            InputStream inStream = GenericConnection.class .getResourceAsStream("jdbc.properties");//获取配置文件的流
            Properties props = new Properties();
            try {
                    props.load(inStream);
                    inStream.close();
            }catch (IOException e1) {
                    e1.printStackTrace();
            }
            class Name = props.getProperty("class Name");
            url = props.getProperty("url");
            uid = props.getProperty("username");
            password = props.getProperty("password");
            try {
                Class .forName(class Name);//加载驱动
            } catch (Class NotFoundException e) {
                e.printStackTrace();
            }
            try {
                //连接数据库
                con = DriverManager.getConnection(url, uid, password);
                System.out.println("数据库连接成功! ");

            } catch (SQLException e) {
                System.out.println("数据库连接失败! ");
                e.printStackTrace();
            }
        }
        return con;
    }
    public static void main(String[] args){
```

```
            getCon();
        }
}
```

连接 MySQL 数据库程序运行结果如图 2-25 所示。

图 2-25　jdbc.properties 文件的位置及连接 MySQL 数据库程序运行结果

2．连接 MySQL 数据库 Python 实现

安装 PyMySQL，如图 2-26 所示。

```
PS C:\Python\Python38-32> pip install PyMySQL
Collecting PyMySQL
  Downloading PyMySQL-1.0.2-py3-none-any.whl (43 kB)
                                                            | 43 kB 70 kB/s
Installing collected packages: PyMySQL
Successfully installed PyMySQL-1.0.2
```

图 2-26　安装 PyMySQL

启动 MySQL，编写以下代码：

```python
import pymysql
# 使用函数装饰器实现单例模式
def singleton(cls):
    _instance = {}
    def inner():
        if cls not in _instance:
            _instance[cls] = cls()
        return _instance[cls]
    return inner
@singleton
class ABC(object):
    def __init__(self):
        pass
    def getMySqlConn(self):
        # 打开数据库连接
        db = pymysql.connect(host="localhost", user="root", passwd="123456")
        # 使用 cursor() 方法创建一个游标对象 cursor
        cursor = db.cursor()
        # 使用 execute() 方法执行 SQL 查询
        cursor.execute("SELECT VERSION()")
        # 使用 fetchone() 方法获取单条数据.
        data = cursor.fetchone()
        print("Database version : %s " % data)
        # 关闭数据库连接
```

```
            db.close()
    a = ABC()
    b = ABC()
    print('a、b的内存地址是否相同：',id(a) == id(b) ,'\na 的内存地址：',id(a) ,'\nb 的
内存地址：', id(b))
    a.getMySqlConn()
    b.getMySqlConn()
```

程序运行结果如图 2-27 所示。

图 2-27　连接 MySQL 数据库 Python 实现运行结果

2.5　多例模式

多例模式（Multiton Pattern）实际上就是单例模式的推广。多例模式分为有上限多例模式和无上限多例模式两种。在有上限多例模式中，多例类的实例是有上限的，当这个多例类的上限数值等于 1 时，多例类退化到单例类。在无上限多例模式中，多例类的实例是没有上限的，也就是说它的上限数值是不确定的。

   应用实例：“华尔街金融网站”支持不同语言

这是一个真实的面向全球消费者的项目的一部分。按照项目计划书，这个网站的系统由数据库驱动，支持 19 种不同的语言，并且将来会支持更多的语言。消费者在登录系统时可以选择自己需要的语言，系统则根据用户的选择将网站的静态文字和动态文字全部转换为用户选择的语言。本节后面会做详细分析并予以实现。

2.5.1　多例模式结构

多例模式和单例模式一般性结构的对比如图 2-28 所示。

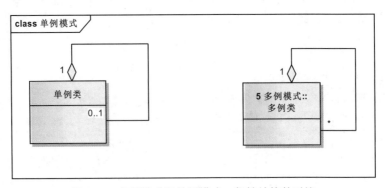

图 2-28　多例模式和单例模式一般性结构的对比

多例模式有以下特点。
（1）多例类可以有多个实例。
（2）多例类必须自己创建自己的实例，管理自己的实例，并向外界提供自己的实例。
（3）多例类分为有上限多例类与无上限多例类。
一个有上限的多例类已经把实例的上限当作逻辑的一部分，并创建在多例类的内部。
具有两个实例的多例类的示意性程序代码如下：

```
package com.multiton;
public class Multiton {
    private static Multiton instance1=null;
    private static Multiton instance2=null;
    private Multiton() { }
    public static Multiton getInstance(int whichOne) {
        if(whichOne==1) {
            if(instance1==null) {
                instance1=new Multiton ();
            }
            return instance1;
        }
        else{
            if(instance2==null){
                instance2=new Multiton ();
            }
            return instance2;
        }
    }
}
```

由于无上限多例模式的多例类对实例的数目是没有限制的，因此，虽然这种多例模式是单例模式的推广，但是这种多例类并不一定能够回到单例类。

多例模式一般采用聚集管理所有的实例。

2.5.2 练习

1．"华尔街金融网站"支持不同语言 Java 实现

经过讨论，设计师们同意对静态文字和动态文字采取不同的解决方案。
（1）把所有的网页交给翻译公司对上面的静态文字进行翻译。
（2）网页上的动态内容需要程序解决。

在进行了研究之后，设计师们发现，他们需要解决的动态文字的"翻译"问题，实际是将数据库中的一些静态或者半静态的数据进行"翻译"。这里形成一个典型的数据表，如表 2-1 所示。

表 2-1 英文数据表

货币代码	货币名称	货币尾数
USD	US,Dollars	2
CNY	China,RenminbiYuan	2
EUR	France,Euro	2
JPY	Japan,Yen	0

在程序中，货币代码不变，货币名称应根据用户选择语言的不同而不同。比如，对中文读者来说，数据表的内容如表 2-2 所示。

表 2-2 中文数据表

货币代码	货币名称	货币尾数
USD	美国（美利坚合众国），美元	2
CNY	中国，人民币元	2
EUR	法国，欧元	2
JPY	日本，日元	0

对日文读者来说，数据表的内容如表 2-3 所示。

表 2-3 日文数据表

货币代码	货币名称	货币尾数
USD	米国（アメリカ合众国），ドルで取引を終えた	2
CNY	中国では，人民元だった	2
EUR	フランスで，ユーロ	2
JPY	日本の円	0

对法文读者来说，数据表的内容如表 2-4 所示。

表 2-4 法文数据表

货币代码	货币名称	货币尾数
USD	Aux États-Unis, dollars	2
CNY	La Chine, le yuan	2
EUR	France, euros	2
JPY	Au Japon, yen	0

这样的表会在网页上作为下拉列表出现，用户看到的是货币名称，而系统内部使用的是货币代码。

系统的内核可以是纯英文的，在内核外部增加一层负责语言翻译工作的翻译层。内核就是系统的模型，而翻译层就是 MVC 模式中的视图层的一部分，支持多种语言的功能属于视图功能。

多例模式把不同的语言转换为不同的参数并产生不同的对象，再将结果传递给读者。

"华尔街金融网站"支持不同的语言，模拟实现结构如图 2-29 所示。

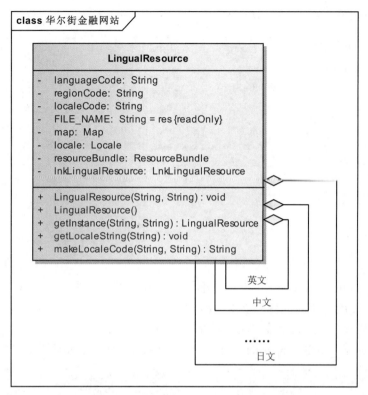

图 2-29 "华尔街金融网站"支持不同的语言模拟实现结构

源代码如下：

```
package com.多例模式.华尔街;

import java.util.HashMap;
import java.util.Locale;
import java.util.Map;
import java.util.ResourceBundle;

public class Multi_Language {
    private static final String FILE_NAME = "res";
    private static Map<Object, Object> map = new HashMap<Object, Object>(10);
    private Locale locale = null;
    private ResourceBundle resourceBundle = null;
    private Multi_Language(String languageCode, String regionCode){
        makeLocaleCode(languageCode , regionCode);
        locale = new Locale(languageCode, regionCode);
        resourceBundle = ResourceBundle.getBundle(FILE_NAME, locale);
    //map.put( makeLocaleCode(languageCode, regionCode) , resourceBundle);
    }
    private Multi_Language(){
    }
```

```
        public synchronized static Multi_Language getInstance(String language,
String region){
            return new Multi_Language( language, region );
        }
        public String getLocaleString(String code) {
            return resourceBundle.getString(code);
        }
        private static String makeLocaleCode(String language, String region){
            return  language + "_" + region;
        }
        public Map getmap(){
            return map;
        }
    }
```

配置文件 res_en_US.properties，该文件在与 com 同级的目录下。这里为了说明 4 种语言都能使用，每条使用对应的语言，单击 "Add" 按钮添加，如图 2-30 所示。注意这里需要用 "UTF-8" 字符编码，具体的设置方法可以在网上自行查找。

配置文件 res_cn_CHN.properties，在实际应用中，每种语言的配置如图 2-31 所示。

图 2-30 配置文件 res_en_US.properties

图 2-31 配置文件 res_cn_CHN.properties

"华尔街金融网站"客户端：

```
package com.multiton.华尔街;
public class Client {
public static void main(String[] args) {
    Multi_Language ling = Multi_Language.getInstance("en", "US");
    showLanguage(ling);
    System.out.println("----------------");
    Multi_Language cling = Multi_Language.getInstance("cn", "CHN");
    showLanguage(cling);
  }
private staticvoid showLanguage(Multi_Language ling){
    String Dollar = ling.getLocaleString("USD");
    System.out.println("USD = " + Dollar);
    String RENMINBIYuan = ling.getLocaleString("CNY");
    System.out.println("CNY = " + RENMINBIYuan);
    String Euro = ling.getLocaleString("EUR");
    System.out.println("EUR = " + Euro);
    String Yen = ling.getLocaleString("JPN");
```

```
            System.out.println("JPN = " + Yen);
    }
}
```

"华尔街金融网站"程序运行结果如图 2-32 所示。

图 2-32　"华尔街金融网站"程序运行结果

在图 2-32 中，分割线上方显示在英文情况下表 2-1 中的参数值，下方显示在中文情况下表 2-2 中的参数值。使用同样方法可以得到其他语言的参数值。

2．Java 线程池/数据库连接池

Java 线程池/数据库连接池是用来管理线程或数据库连接对象的，其用途一是限制池中对象的数量，二是能够在使用过程中达到复用的效果。

线程池中的线程在任务执行完毕后，不会被直接回收，而是切换成等待状态，等待下一个任务的提交、执行。数据库连接池也是如此，数据库在连接操作完毕后，会把资源释放回连接池，然后等待下一次数据库连接操作。

这种设计其实是将对象的应用最大化了，避免了在每次连接的时候都需要创建一个对象，造成对象冗余或内存升高。

```
package com.多例模式.线程池;

import java.util.ArrayList;
import java.util.List;
import java.util.Random;
public class SQLConnectionPools {
    private static int maxNumOfConnection= 3;
    private static List<String> connectionInfoList = new ArrayList<String>(maxNumOfConnection);
    private static List<SQLConnectionPools> connArrayList = new ArrayList<SQLConnectionPools>(maxNumOfConnection);
    private static int currNumOfConnection =0;
    private SQLConnectionPools() {
        // TODO Auto-generated constructor stub
    }
    private SQLConnectionPools(String info) {
        connectionInfoList.add(info);
    }
    static{
        for (int i = 0; i < maxNumOfConnection; i++) {
```

```java
            connArrayList.add(new SQLConnectionPools(i+"号连接"));
        }
    }
    public static SQLConnectionPools getInstance() {
        Random random = new Random();
        currNumOfConnection = random.nextInt(maxNumOfConnection);
        return connArrayList.get(currNumOfConnection);
    }

    public void connectionInfo() {
        System.out.println(connectionInfoList.get(currNumOfConnection));
    }
}
```

测试程序如下:

```java
package com.多例模式.线程池;
public class Client {
    public static void main(String[] args) {
        // TODO Auto-generated method stub
        int userNum=20;
        for(int i=0;i<userNum;i++){
            //用户获取到的连接是随机的
            SQLConnectionPools conn= SQLConnectionPools.getInstance();
            System.out.print("第"+i+"个用户获得的连接是: ");
            conn.connectionInfo();
        }
    }
}
```

程序运行结果如图 2-33 所示。

```
<terminated> Client (4) [Java Ap
第0个用户获得的连接是: 1号连接
第1个用户获得的连接是: 1号连接
第2个用户获得的连接是: 1号连接
第3个用户获得的连接是: 1号连接
第4个用户获得的连接是: 2号连接
第5个用户获得的连接是: 0号连接
第6个用户获得的连接是: 0号连接
第7个用户获得的连接是: 1号连接
第8个用户获得的连接是: 2号连接
第9个用户获得的连接是: 2号连接
第10个用户获得的连接是: 0号连接
第11个用户获得的连接是: 1号连接
第12个用户获得的连接是: 1号连接
第13个用户获得的连接是: 1号连接
第14个用户获得的连接是: 2号连接
第15个用户获得的连接是: 2号连接
第16个用户获得的连接是: 0号连接
第17个用户获得的连接是: 2号连接
第18个用户获得的连接是: 0号连接
第19个用户获得的连接是: 0号连接
```

图 2-33　Java 线程池/数据库连接池系统运行结果

3. 多例模式 Python 实现

Python 默认的都是多例模式,验证代码如下:

```
class custom():
    pass
c1 = custom()
c2 = custom()
print(c1)
print(c2)
# 我们可以通过它们的内存地址来判断是否是同一个实例,也可以使用 is 判断
print("c1 和 c2 是否是同一个实例:",c1 is c2)
```

程序运行结果如图 2-34 所示。

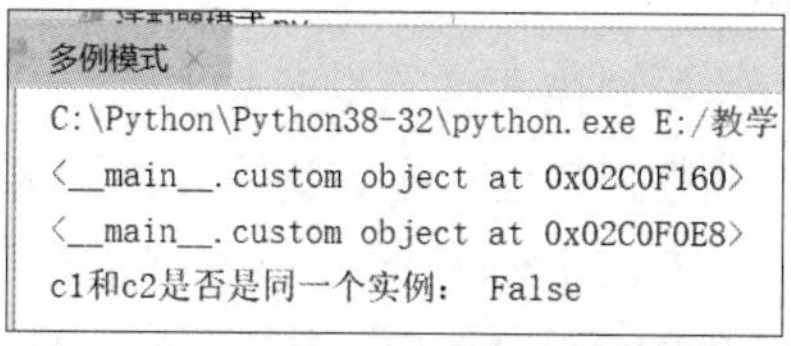

```
多例模式 ×
C:\Python\Python38-32\python.exe E:/教学
<__main__.custom object at 0x02C0F160>
<__main__.custom object at 0x02C0F0E8>
c1和c2是否是同一个实例: False
```

图 2-34 多例模式 Python 实现程序运行结果

第3章

结构型模式

结构型模式（Structural Pattern）描述如何将类或对象结合在一起以形成更大的结构。结构型模式描述两种不同的东西：类与类的对象。根据这一不同，结构型模式可以分为类的结构型模式和对象的结构型模式两种。

（1）类的结构型模式：类的结构型模式使用继承把类、接口等组合在一起，以形成更大的结构。当一个类从父类继承并实现某接口时，这个新的类就把父类的结构和接口的结构结合起来。类的结构型模式是静态的。类的结构型模式的典型例子就是适配器模式。

（2）对象的结构型模式：对象的结构型模式描述怎样把各种不同类型的对象组合在一起，以实现新的功能。对象的结构型模式是动态的。对象的结构型模式的典型例子就是代理人模式，其他的例子包括后面将要介绍的装饰模式以及对象形式的适配器模式。

本章将要介绍的结构型模式包括适配器模式、默认适配器模式、装饰模式和门面模式，有一些模式会有类形式和对象形式两种。下面要介绍的适配器模式就是这样，它有类形式和对象形式两种。

3.1 适配器模式

适配器模式（Adapter Pattern）把一个类的接口转换为客户端期待的另一种接口，从而使因接口不匹配而无法在一起工作的两个类能够在一起工作。适配器类可以根据参数给客户端返还一个合适的实例。

意图：将一个类的接口转换为客户希望的另外一个接口。适配器模式使原本由于接口不兼容而不能一起工作的类可以一起工作。

主要解决：主要解决在软件系统中将一些"现存的对象"放到新的环境中，而新环境要求的接口是现对象不能满足的情况。

> **应用实例：手机充电器**

美国民用电器的额定电压是110V，中国民用电器的额定电压是220V，因此，需要有一个适配器将110V转换为220V。

优点：适配器模式可以让任何两个没有关联的类一起运行，提高了类的复用，增加了类的透明度，灵活性好。

缺点：过多地使用适配器会让系统非常零乱，不易整体把握。

使用场景：有动机地修改一个正常运行的系统的接口。

注意事项：适配器不是在详细设计时添加的，适配器是为了解决正在使用中的项目出现的新问题。

3.1.1 适配器模式的结构

适配器模式有类适配器模式和对象适配器模式两种。

1. 类适配器模式

类适配器模式把适配的类的 API 转换为目标类的 API，其结构如图 3-1 所示。

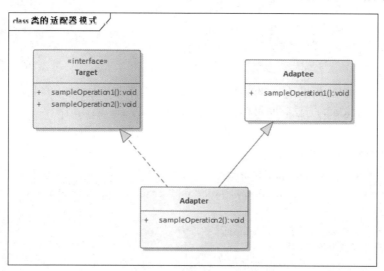

图 3-1 类适配器模式的结构

由上图可以看出，Adaptee 类没有 sampleOperation2()方法，而客户端需要这个方法。

为了使客户端能够使用 Adaptee 类，提供一个中间环节即 Adapter 类，把 Adaptee 类的 API 与 Target 类的 API 连接起来。

Adapter 类与 Adaptee 类是继承关系，因此，这个适配器模式是类适配器模式。

该模式所涉及的角色有以下 3 种。

（1）源（Adaptee）角色：需要适配的接口。

（2）目标（Target）角色：期待得到的接口。

（3）适配器（Adapter）角色：适配器类是适配器模式的核心。适配器把源角色转换为目标角色。适配器必须是具体类。

类适配器模式源代码如下。

目标（Target）角色：

```
package com.adapter.class Adapter;
public interface Target {
    /**
     * 这是源类Adaptee也有的方法
     */
    public void sampleOperation1();
    /**
     * 这是源类Adaptee没有的方法
```

```
     */
    public void sampleOperation2();
}
```

目标角色是以一个 Java 接口的形式实现的。可以看出，这个接口声明了两个方法：sampleOperation1()和 sampleOperation2()。而源角色 Adaptee 是一个具体类，它有一个 sampleOperation1()方法，但是没有 sampleOperation2()方法。

源（Adaptee）角色：

```
package com.adapter.class Adapter;
public abstract class Adaptee {
    public void sampleOperation1(){}
}
```

适配器角色 Adapter 扩展了 Adaptee，同时又实现了目标（Target）接口。由于 Adaptee 没有提供 sampleOperation2()方法，而目标接口又要求这个方法，因此适配器角色 Adapter 实现了这个方法。

适配器（Adapter）角色：

```
package com.adapter.class Adapter;
public class Adapter extends Adaptee implements Target {
    /**
     * 由于源类Adaptee没有方法sampleOperation2()
     * 因此适配器补充这个方法
     */
    public void sampleOperation2() {
        //写相关的代码
    }
}
```

2．对象适配器模式

与类适配器模式一样，对象适配器模式把被适配的类的 API 转换为目标类的 API。与类适配器模式不同的是，对象适配器模式不是使用继承关系连接 Adaptee 类的，而是使用委派关系连接 Adaptee 类的。对象适配器模式的结构如图 3-2 所示。

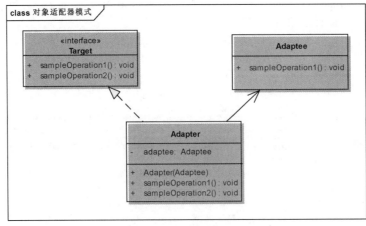

图 3-2　对象适配器模式的结构

从上图可以看出，Adaptee 类并没有 sampleOperation2() 方法，但客户端期待这个方法。为了使客户端能够使用 Adaptee 类，需要提供一个包装（Decorator）类 Adapter。这个包装类包装了一个 Adaptee 的实例，能够把 Adaptee 类的 API 与 Target 类的 API 连接起来。Adapter 类与 Adaptee 类是委派关系，因此，这个适配器模式是对象的适配器模式。

对象适配器模式源代码如下。

目标（Target）角色：

```java
package com.adapter.objectAdapter;

public interface Target {
    /**
     * 这是源类 Adaptee 也有的方法
     */
    public void sampleOperation1();
    /**
     * 这是源类 Adaptee 没有的方法
     */
    public void sampleOperation2();
}
```

源（Adaptee）角色：

```java
package com.adapter.objectAdapter;
public abstract class Adaptee {
    public void sampleOperation1(){}
}
```

适配器（Adapter）角色：

```java
package com.adapter.objectAdapter;

public class Adapter implements Target{
    private Adaptee adaptee;
    public Adapter(Adaptee adaptee){
        this.adaptee = adaptee;
    }
    /**
     * 源类 Adaptee 有方法 sampleOperation1
     * 因此适配器类直接委派即可
     */
    public void sampleOperation1(){
        this.adaptee.sampleOperation1();
    }
    /**
     * 源类 Adaptee 没有方法 sampleOperation2
     * 因此，适配器类需要补充此方法
```

```
     */
    public void sampleOperation2(){
        //写相关的代码
    }
}
```

3. 类适配器和对象适配器的权衡

(1) 类适配器使用对象继承的方式,是静态的定义方式。对象适配器使用对象组合的方式,是动态的组合方式。

(2) 对于类适配器,因为继承是静态的关系,当适配器继承了 Adaptee 后,就不可能再去处理 Adaptee 的子类了。对于对象适配器,一个适配器可以把多种不同的源适配到同一个目标。换言之,同一个适配器可以把源类和它的子类都适配到目标接口。因为对象适配器采用的是对象组合的关系,只要对象类型正确,是不是子类都无所谓。

(3) 类适配器可以重定义 Adaptee 的部分行为,相当于子类覆盖父类的部分实现方法。对象适配器重定义 Adaptee 的行为比较困难,在这种情况下,需要定义 Adaptee 的子类来实现重定义,然后让适配器组合子类。虽然重定义 Adaptee 的行为比较困难,但是想要增加一些新的行为很方便,而且新增加的行为可以同时适用于所有的源。

(4) 类适配器仅仅引入了一个对象,并不需要额外的引用来间接得到 Adaptee。对象适配器需要额外的引用来间接得到 Adaptee。

软件设计原则建议尽量使用对象适配器的实现方式,多用合成/聚合,少用继承。当然具体问题需要具体分析,最适合的才是最好的。

3.1.2 电源适配器实现

目标(Target)角色:

```java
package com.adapter;
/**
 *目标角色
 */
public interface Target {
    int get110v();
    int get220v();
}
```

源(Adaptee)角色:

```java
package com.adapter;
/**
 *需要适配的类
 */
public class Adaptee {
    public int get110v(){
        return 110;
    }
}
```

类适配器（Adapter）角色：

```java
package com.adapter;
/**
 * 类适配器角色可以扩展源角色，实现目标角色，从而在改动目标角色的时候，不用改动源角色，只
需要改动适配器
 */
public class Adapter extends Adaptee implements Target{
    public int get220v(){
        Adaptee adaptee = new Adaptee();
        //这里是转换函数
        int t = adaptee.get110v();
        t = t*2;
        return t;
    }
}
```

对象适配器（Adapter）角色：

```java
package com.adapter;
//对象适配器
public class OAdapter implements Target{
    Adaptee adaptee;
    public OAdapter(Adaptee adaptee) {
        super();
        this.adaptee = adaptee;
    }
    public int get110v(){
        return adaptee.get110v();
    }

    public int get220v(){
        Adaptee adaptee = new Adaptee();

        //这里是转换函数
        int t = adaptee.get110v();
        t = t*2;

        return t;
    }
}
```

类客户端实现：

```java
package com.adapter;
/**
```

```java
* 测试类适配器
*/
public class Client {
public static void main(String rags[]) {
new Client().test();
    }
public void test() {
    Target target = newAdapter();
//转换前
int v1 = target.get110v();
    System.out.println("转换前:" + v1);
//转换后
int v2 = target.get220v();
    System.out.println("转换后: " + v2);
  }
}
```

对象客户端实现：

```java
package com.adapter;
//测试对象适配器
public class OClient {
    public static void main(String[] args) {
        new Client().test();

    }
public void test() {
    Adaptee adaptee = new Adaptee();
    Target target = new OAdapter(adaptee);
//转换前
int v1 = target.get110v();
    System.out.println("转换前:" + v1);
//转换后
int v2 = target.get220v();
    System.out.println("转换后: " + v2);
    }
}
```

两种适配器程序运行结果相同，如图3-3所示。

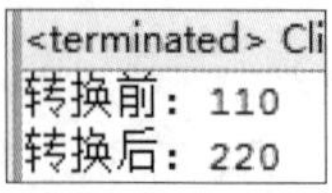

3.1.3 适配器模式的优缺点

1. 适配器模式的优点

图3-3 适配器程序运行结果

（1）更好的复用性。如果系统需要使用现有的类，但类的接口不符合系统的需要，那么通过适配器模式就可以让这些功能得到更好的复用。

（2）更好的扩展性。在实现适配器功能的时候，可以调用自己开发的功能，从而自然地扩展系统的功能。

2. 适配器模式的缺点

过多地使用适配器，会让系统非常零乱，不易整体把握。比如，明明看到调用的是 A 接口，其实内部被适配成了 B 接口。一个系统如果过多地出现这种情况，无异于一场灾难。因此，如果不是很有必要，可以不使用适配器，而是直接对系统进行重构。

3.1.4 练习

1. 手机充电器 Java 实现

充电器本身相当于 Adapter，220V 交流电相当于 Adaptee（被适配者），Target 是 5V 直流电。按此思路设计的系统结构如图 3-4 所示。

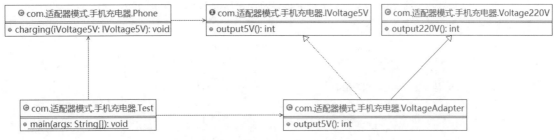

图 3-4 使用适配器模式的手机充电器的系统结构

代码实现如下：

```
//被适配的类，即手机充电器的输入电压
package com.适配器模式.手机充电器;
public class Voltage220V {
    public int output220V() {
        int src = 220;
        System.out.println("可提供的电压为" + src + "伏");
        return src;
    }
}
//适配接口，即手机充电器的输出电压
package com.适配器模式.手机充电器;
public interface IVoltage5V {
    public int output5V();
}
//适配器类，即手机充电器
package com.适配器模式.手机充电器;
public class VoltageAdapter extends Voltage220V implements IVoltage5V {

    public int output5V() {
        //获取 220V 电压
```

```java
        int srcV = output220V();
        int dstV = srcV / 44 ;  //转换为5v
        return dstV;
    }
}
package com.适配器模式.手机充电器;
public class Phone {
    //充电
    public void charging(IVoltage5V iVoltage5V) {
        if(iVoltage5V.output5V() == 5) {
            System.out.println("充电器输出电压为5V,可以充电~~");
        } else if (iVoltage5V.output5V() > 5) {
            System.out.println("充电器输出电压大于5V,不能充电~~");
        }
    }
}
package com.适配器模式.手机充电器;
public class Test {
    /**
     * @param args
     */
    public static void main(String[] args) {
        // TODO Auto-generated method stub
        System.out.println(" === 类适配器模式 ====");
        Phone phone = new Phone();
        phone.charging(new VoltageAdapter());
    }
}
```

程序运行结果如图3-5所示。

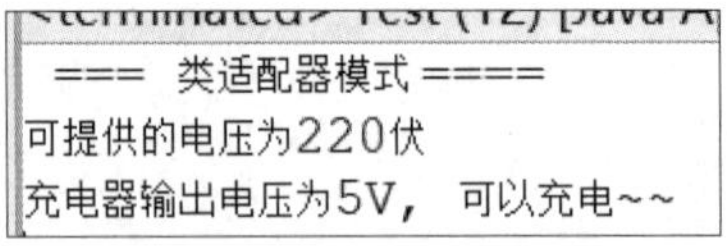

```
=== 类适配器模式 ====
可提供的电压为220伏
充电器输出电压为5V, 可以充电~~
```

图3-5 手机充电器Java实现程序运行结果

2. 适配器模式Python实现

使用Python实现适配器模式有多种方法，比如继承、组合，但是Python提供了一个替代方案，即使用类的内部字典。

下面使用内部字典实现适配器模式。

```python
# 源(Adaptee)角色:
class Computer:
    def __init__(self,name):
```

```python
        self.name=name
    def __str__(self):
        return 'the {} computer'.format(self.name)
    def execute(self):
        return 'executes a program'

# 目标（Target）角色：
class Synthesizer:
    def __init__(self, name):
        self.name = name

    def __str__(self):
        return 'the {} synthesizer'.format(self.name)

    def play(self):
        return 'is playing an electronic song'

# 目标（Target）角色：
class Human:
    def __init__(self, name):
        self.name = name
    def __str__(self):
        return '{} the human'.format(self.name)
    def speak(self):
        return 'says hello'
```

客户端仅需要知道如何调用 execute()方法，不需要知道 play()和 speak()。创建的 Adapter 类将带有不同接口的对象适配到一个接口中。

__init__方法的 obj 参数是我们想要适配的对象，adapted_methods 是一个字典，键值对中的键是客户端要调用的方法，值是应该被调用的方法。

```python
# 适配器（Adapter）角色：
class Adapter:
    def __init__(self, obj, adapted_methods):
        self.obj = obj
        self.__dict__.update(adapted_methods)
    def __str__(self):
        return str(self.obj)

if __name__=="__main__":
    objects=[Computer('Asus')]
    print(objects[0].execute())
```

```python
synth=Synthesizer('moog')
objects.append(Adapter(synth,dict(execute=synth.play)))
print('不使用适配器模式: ',synth.play())
print('使用适配器模式: ',objects[1].execute())
human=Human('Bob')
objects.append(Adapter(human,dict(execute=human.speak)))
print(objects[2].execute())
for i in objects:
    print('{} {}'.format(str(i),i.execute()))
```

程序运行结果如图 3-6 所示。

图 3-6 适配器模式 Python 实现程序运行结果

3.2 默认适配器模式

默认适配器模式（Default Adapter Pattern）为一个接口提供默认实现，在这之后，类型可以从这个默认实现进行扩展，而不必从原有的接口进行扩展。作为适配器模式的一个特例，默认适配器模式在 Java 语言中有着特殊的应用。

应用实例：鲁智深受戒

和尚要做什么呢？吃斋、念经、打坐、撞钟、习武等。按照常规的做法，鲁智深想成为一个和尚应该如图 3-7 所示。

但鲁智深除了会武术，其他的都不会，只能实现 getName()和习武()方法，不能实现其他的方法。因此，它根本就通不过 Java 语言编译器。鲁智深类只有实现和尚接口的所有的方法才可以通过 Java 语言编译器，但是这样一来鲁智深就不再是鲁智深了。

研究一下几百年前鲁智深是怎么剃度成为和尚的，会对学习 Java 编程有很大的启发。当初鲁达剃度，众僧说："此人形容丑恶，相貌凶顽，不可剃度他。"但是长老却说："此人上应天星，心地刚直。虽然时下凶顽，命中驳杂，久后却得清净，正果非凡，汝等皆不及他。"

原来如此！看来只要这里应上一个天星，问题就解决了！对面向对象的语言来说，"应"者，实现也。"天星"者，抽象类也。

采用默认适配器模式后，鲁智深也就成了和尚，设计方案如图 3-8 所示。

第 3 章 结构型模式

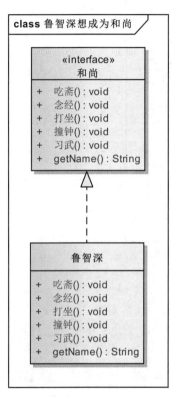
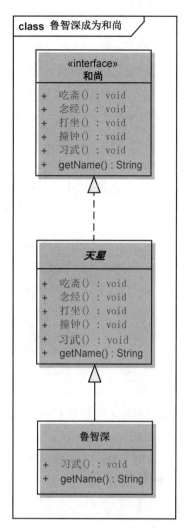

图 3-7 在常规情况下鲁智深想成为和尚设计方案　图 3-8 采用默认适配器模式后鲁智深成为和尚设计方案

3.2.1 默认适配器模式的结构

在很多情况下，必须让一个具体类实现某一个接口，但是这个类又用不到接口规定的所有的方法。通常的处理方法是，这个具体类要实现所有的方法，有用的方法要有实现，没有用的方法也要有空的实现。

这些空的方法是一种浪费，有时也是一种混乱。如果没有看过这些空方法的代码，程序员可能会以为这些方法不是空的。除非看过这些方法的源代码或是文档，否则，即便他知道其中有一些方法是空的，也不一定知道哪些方法是空的，哪些方法不是空的。

默认适配器模式是一种"平庸化"的适配器模式。

默认适配器模式可以很好地处理这一情况。默认适配器模式可以设计一个抽象的适配器类实现接口，此抽象类要给接口要求的每一种方法都提供一个空的方法。就像帮助了鲁智深的"天星"一样，此抽象类可以使具体类的子类免于被迫实现空的方法。

默认适配器模式的结构设计如图 3-9 所示。

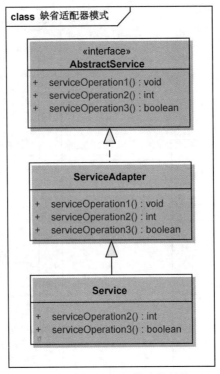

图 3-9 默认适配器模式的结构

默认适配器模式示意性源代码如下。

接口 AbstractService 类：

```java
package com.defaultAdapter;
public interface AbstractService {
    public void    serviceOperation1();
    public int     serviceOperation2();
    public boolean serviceOperation3();
}
```

抽象适配器 ServiceAdapter 类：

```java
package com.defaultAdapter;
public abstract class  ServiceAdapter implements AbstractService {
    public void  serviceOperation1() {
    }
    public abstract int serviceOperation2();
    public abstract boolean serviceOperation3();
}
```

具体类 Service 继承抽象类 ServiceAdapter：

```java
package com.defaultAdapter;
public class  Service  extends  ServiceAdapter {
    public int serviceOperation2() {
        // 实现所需的方法
        return  0;
```

```
        }
    public boolean serviceOperation3() {
            //实现所需的方法
            return false;
        }
}
```

可以看到，接口 AbstractService 要求定义 3 种方法，分别是 serviceOperation1()、serviceOperation2()、serviceOperation3()。抽象适配器类 ServiceAdapter 为这 3 种方法都提供了平庸的实现。因此，继承抽象类 ServiceAdapter 的具体类 Service 可以选择它需要的方法实现，不必理会其他的不需要的方法。

适配器模式的用意是改变源的接口，以便与目标接口相容。默认适配器模式的用意稍有不同，它是为了方便建立一个不平庸的适配器类而提供的一种平庸实现。

在任何时候，如果不准备实现一个接口的所有方法，就可以使用默认适配器模式创建一个抽象类，给出所有方法的平庸的具体实现。这样，子类就不必实现从这个抽象类继承的所有的方法了。

3.2.2 练习

1. "鲁智深受戒" Java 实现

采用默认适配器模式后鲁智深成为和尚的实现如下。

源代码如下。

和尚接口，给出所有的和尚都需要实现的方法：

```
package com.defaultAdapter.monk;
public interface 和尚 {
    public void 吃斋();
    public void 念经();
    public void 打坐();
    public void 撞钟();
    public void 习武();
    public String getName();
}
```

抽象类作为适配器类：

```
package com.defaultAdapter.monk;
public abstract class 天星 implements 和尚 {
    public void 吃斋(){}
    public void 念经(){}
    public void 打坐(){}
    public void 撞钟(){}
    public abstract void 习武();
    public abstract String getName();
}
```

鲁智深类继承抽象类"天星":

```
package com.defaultAdapter.monk;
public class 鲁智深 extends 天星{
  public void 习武(char ch){
    Character c = null;
    ch = c.toUpperCase(ch);
    switch(ch){
    case'A': {
        System.out.println("拳打镇关西");
        break;
    }
    case'B':
        System.out.println("大闹五台山");
        break;
    case'C':
        System.out.println("大闹桃花村");
        break;
    case'D':
        System.out.println("火烧瓦罐寺");
        break;
    case'E':
        System.out.println("倒拔垂杨柳");
        break;
    default: {
        System.out.println("退出，再见！");
        System.exit(0);
        break;
    }
    }
  }
  public String getName(){
        return "我是鲁智深";
  }
}
```

客户端:

```
package com.defaultAdapter.monk;

import Java.util.Scanner;

public class  Client {
    public static void main(String[] args) {
```

```
            String i;
            char ch;
            鲁智深 鲁智深 = new 鲁智深();
            System.out.println(鲁智深.getName());
            System.out.println("请输入 A-E，不分大小写，其他字符退出：");
            Scanner sc;
    while(true){
            sc = new Scanner(System.in);
            i = sc.next();
            ch = i.charAt(0);
            鲁智深.习武(ch);
        }
    }
}
```

程序运行结果如图 3-10 所示。

图 3-10 "鲁智深受戒" Java 实现程序运行结果

鲁智深借助适配器模式达到了剃度的目的。此适配器类实现了和尚接口要求的所有方法。但是与通常的适配器模式不同的是，此适配器类给出的所有的方法的实现都是"平庸"的。这种"平庸化"的适配器模式称为默认适配器模式。

2. 手机充电接口 Python 实现

```
#target, 需要安卓的充电线
class Android:
    def __init__(self):
    pass

    def connect_mobile(self, mobile):
    print('我是安卓手机，需要安卓手机充电线')
        mobile.connectAn()



    # adaptee, 这是个苹果手机的充电线
```

```python
class Apple:
    def __init__(self):
        pass

    def connectAp(self):
        print('有苹果手机充电线')

# adapter
class Adapter:
    def __init__(self):
        self._connectApple = Apple()

    def connectAn(self):
        print('适配器来了，是个转换接口，苹果充电线接上适配器就能给安卓手机充电了')
        self._connectApple.connectAp()

if __name__ == '__main__':
    androidM = Android()
    adapter = Adapter()
    androidM.connect_mobile(adapter)
```

程序运行结果如图 3-11 所示。

图 3-11 手机充电接口 Python 实现程序运行结果

3.3 装饰模式

装饰模式（Decorator Pattern）又名包装模式。装饰模式以对客户端透明的方式扩展对象的功能，是继承关系的一个替代方案，比继承关系更灵活。装饰模式可以动态地给一个对象增加功能，这些功能可以再动态地撤销，还可以增加由一些基本功能的排列组合而产生的功能。

3.3.1 应用实例：孙悟空七十二般变化

孙悟空有七十二般变化，他的每一种变化都给他带来一种附加的本领。他变成鱼儿时，就可以到水里游泳；他变成鸟儿时，就可以在天上飞行。使用装饰模式的"孙悟空七十二般变化"的设计方案如图 3-12 所示。

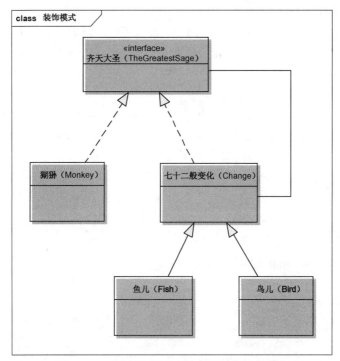

图 3-12 "孙悟空七十二般变化"的设计方案

3.3.2 装饰模式的结构

装饰模式以对客户透明的方式动态地给一个对象附加更多的责任。换言之，客户端并不会觉得对象在装饰前和装饰后有什么不同。装饰模式可以在不创建更多子类的情况下，扩展对象的功能。

装饰模式的结构如图 3-13 所示。

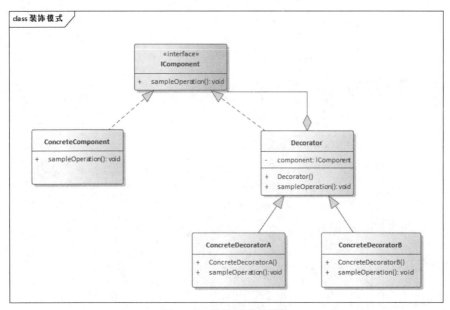

图 3-13 装饰模式的结构

在装饰模式中的角色有以下几种。

（1）抽象构件（Component）角色：给出一个抽象接口，以规范准备接收附加责任的对象。

（2）具体构件（ConcreteComponent）角色：定义一个将要接收附加责任的类。

（3）装饰（Decorator）角色：持有一个构件（Component）对象的实例，并定义一个与抽象构件接口一致的接口。

（4）具体装饰（ConcreteDecorator）角色：负责给构件对象"贴上"附加的责任。

源代码如下。

抽象构件（Component）角色：

```java
package com.decorator;
public interface IComponent {
    public void sampleOperation ();
}
```

具体构件（ConcreteComponent）角色：

```java
package com.decorator;
public class ConcreteComponent implements IComponent {
    public ConcreteComponent(){}
    public void sampleOperation() {
        System.out.println("开车");
    }
}
```

装饰（Decorator）角色：

```java
package com.decorator;
public class Decorator implements IComponent {
    private IComponent component;
    public Decorator(IComponent component) {
        this.component=component;
    }
    public void sampleOperation() {
        component.sampleOperation();
    }
}
```

具体装饰（ConcreteDecorator）角色 A：

```java
package com.decorator;
public class ConcreteDecoratorA extends Decorator{
public ConcreteDecoratorA(IComponent component){
        super(component);
    }
    public void sampleOperation(){
        this.addedOperation();
        super.sampleOperation();
    }
```

```
        private void addedOperation() {
            System.out.println("晚上");
        }
}
```

具体装饰（ConcreteDecorator）角色 B：

```
package com.decorator;
public class ConcreteDecoratorB extends Decorator{
    public ConcreteDecoratorB(IComponent component){
        super(component);
    }
    public void sampleOperation(){
        this.addedOperation();
        super.sampleOperation();
    }
    private void addedOperation() {
        System.out.println("早上");
    }
}
```

客户端：

```
package com.decorator;
public class Client {
    public static void main(String[] args) {
        IComponent component = new ConcreteComponent();

        Decorator decorator = new ConcreteDecoratorA(component);
        decorator.sampleOperation();
        decorator = new ConcreteDecoratorB(component);
        decorator.sampleOperation();
    }
}
```

程序运行结果如图 3-14 所示。

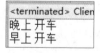

图 3-14 程序运行结果

3.3.3 "孙悟空七十二般变化" Java 实现

本例中，IComponent 的角色就是由大名鼎鼎的齐天大圣扮演的。ConcreteComponent 的角色就是大圣的本尊。Decorator 的角色由大圣的"七十二变"扮演。而 ConcreteDecorator 的角色就是"鱼儿""鸟儿"等七十二般变化。

"孙悟空七十二般变化"的 Java 实现源代码如下。

抽象构件角色"齐天大圣"接口：定义了一个 move()方法，这是所有的具体构件类和装饰类必须实现的。

```
package com.decorator.monkey;
public interface I齐天大圣 {
    public void I齐天大圣();
    public void move();
}
```

具体构件角色"大圣本尊"，即"猢狲"：

```
package com.decorator.monkey;
public class 大圣本尊 implements I齐天大圣 {
    public 大圣本尊(){
        System.out.println("我就是大圣本尊！");
    }
    public void move() {
        System.out.println("我来也！");
    }
    public void I齐天大圣() {
        System.out.println("我是齐天大圣孙悟空！");
    }
}
```

抽象装饰角色"七十二变"：

```
package com.decorator.monkey;
public abstract class 七十二变 {
    private I齐天大圣 c;
    public 七十二变(I齐天大圣 c){
        this.c = c;
    }
    public void move(){
        c.move();
    }
    abstractpublic void 外形();
}
```

具体装饰角色"鱼儿"：

```
package com.decorator.monkey;
public class 鱼 extends 七十二变 {
    public 鱼(I齐天大圣 c) {
        Super(c);
        外形();
    }
```

```java
    public void move(){
        System.out.print("在水中,");
        super.move();
    }
    public void 外形() {
        System.out.println("哈哈,看我变成一条鱼!");
    }
}
```

具体装饰角色"鸟儿":

```java
package com.decorator.monkey;
public class 鸟 extends 七十二变 {
    public 鸟(I齐天大圣 c) {
        super(c);
        外形();
    }
    public void move(){
        System.out.print("在空中,");
        super.move();
    }
    public void 外形() {
        System.out.println("哈哈,看我变成一只鸟!");
    }
}
```

客户端:

```java
package com.decorator.monkey;
public class Client{
    public static void main(String[] args) {
        I齐天大圣  monkeyKing;
        monkeyKing = new 大圣本尊();
        monkeyKing.move();
        
        七十二变 change ;
        change = new 鱼(monkeyKing);
        change.move();
        change = new 鸟(monkeyKing);
        change.move();
    }
}
```

程序运行结果如图 3-15 所示。

图 3-15 "孙悟空七十二般变化" Java 实现程序运行结果

"大圣本尊"是 ConcreteComponent 类,而"鸟儿""鱼儿"是装饰类。要装饰的是"大圣本尊",即"猢狲"实例。

3.3.4 装饰模式的简化

在大多数情况下,装饰模式的实现都要比前一节给出的例子简单。

如果只有一个 ConcreteComponent 类,那么可以考虑去掉抽象的 IComponent 类(接口),把 Decorator 作为 ConcreteComponent 的子类,这种简化的装饰模式的结构如图 3-16 所示。

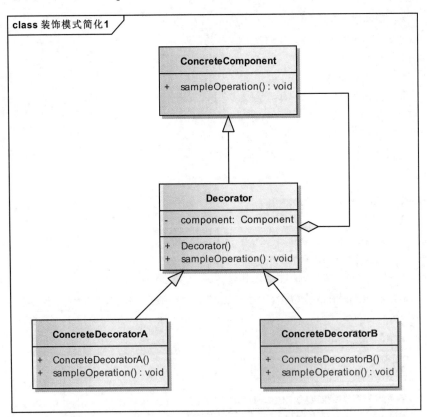

图 3-16 简化的装饰模式的结构

如果只有一个 ConcreteDecorator 类,就没有必要建立一个单独的 Decorator 类,在这种情况下,可以把 Decorator 和 ConcreteDecorator 的责任合并成一个类。在只有两个 ConcreteDecorator 类的情况下,也可以这样做,进一步简化的装饰模式的结构如图 3-17 所示。

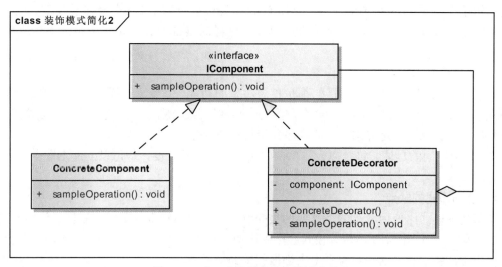

图 3-17 进一步简化的装饰模式的结构

3.3.5 装饰模式进阶

1. 透明性的要求

装饰模式对客户端的透明性要求程序不要声明一个 ConcreteComponent 类型的变量，而应当声明一个 IComponent 类型的变量。

用孙悟空的例子来说，必须永远把孙悟空的所有变化都当成孙悟空来对待，如果把孙悟空变成的鱼儿当成鱼儿而不是孙悟空，就被孙悟空骗了，这是不应当发生的。下面的做法是对的：

```
TheGreatestSage sage = new Monkey();
TheGreatestSage bird = new Bird(sage);
```

而下面的做法是不对的，违反了依赖倒置原则。

```
Monkey sage = new Monkey();
Bird bird = new Bird(sage);
```

2. 半透明的装饰模式

纯粹的装饰模式是很难找到的。装饰模式的用意是在不改变接口的前提下，增强考虑的类的性能。在增强性能的时候，往往需要建立新的公开的方法。即便是在孙悟空的系统里，也需要新的方法。比如，齐天大圣类并没有飞行的能力，而鸟儿有，这就意味着鸟儿应当有一个新的 fly() 方法。再比如，齐天大圣类并没有游泳的能力，而鱼儿有，这就意味着在鱼儿类里应当有一个新的 swim() 方法。

这就导致了大多数的装饰模式的实现都是半透明的，而不是完全透明的。换言之，允许装饰模式改变接口，增加新的方法。这意味着客户端可以声明 ConcreteDecorator 类型的变量，从而调用 ConcreteDecorator 类中才有的方法。

```
TheGreatestSage sage = new Monkey();
Bird bird = new Bird(sage);
bird.fly();
```

半透明的装饰模式是介于装饰模式和适配器模式之间的一种模式。适配器模式的用意是改变类的接口,也可以通过改写一个或几个方法,或增加新的方法来增强或改变类的功能。大多数的装饰模式实际上是半透明的装饰模式,这样的装饰模式也称作半装饰模式或半适配器模式。

3．装饰模式的优点

(1)装饰模式与继承关系的目的都是扩展对象的功能,但是装饰模式比继承关系更灵活。装饰模式允许系统动态地"贴上"一个需要的"装饰",或者删除一个不需要的"装饰"。继承关系则不同,继承关系是静态的,它在系统运行前就决定了。

(2)使用不同的具体装饰类以及这些装饰类的排列组合,设计师可以创造出很多不同行为的组合。

4．装饰模式的缺点

使用装饰模式比使用继承关系需要的类的数目少。使用较少的类,可以使设计容易进行。但是,从另一方面来说,使用装饰模式会产生比使用继承关系更多的对象。更多的对象会使查错变得困难,特别是在这些对象看上去都很相像的情况下。

3.3.6 练习

1．电子销售系统打印发票构件 Java 实现

1)需求分析

一个电子销售系统需要打印出顾客购买商品的发票。一张发票可以分为 3 个部分。

(1)发票头部(Header):上面有顾客的名字和销售的日期。

(2)发票的主体部分:销售的货物清单,包括商品名字、购买的数量、单价、小计。

(3)发票的尾部(Footer):商品的总金额。

发票的头部和尾部可以有很多种可能的格式,客户端必须可以随意地选择某一个头部格式和某一个尾部格式的组合并与发票主体部分的格式结合起来。

2)系统设计

为了解决面临的问题,必须选择一个合适的设计,使系统可以处理很多的头部格式和尾部格式的组合,而这就暗示着设计师会考虑使用装饰模式。装饰模式可以将更多的功能动态地附加到一个对象上。对功能扩展而言,装饰模式提供了一个灵活的、可以替代继承的选择。

如果在这个系统里使用装饰模式,那么发票的头部和尾部可以分别由具体装饰类 HeaderDecorator 和具体装饰类 FooterDecorator 代表。有多少种头部和尾部,就可以有多少种相应的具体装饰类。这样一来,就可以选择大量的头部和尾部的组合。

该系统的设计方案如图 3-18 所示。

在电子销售系统设计方案中,各部分代表的角色或提供的功能如下。

Order 是抽象构件角色。

SalesOrder 代表发票的主体部分。

OrderDecoratcr 是抽象装饰角色。

OrderLine 是一张发票的货品清单中的一行,给出产品名、产品单价、所购单位数、小计金额等。

Order 类有一个 Vector 聚集,用以存储多个 OrderLine 对象。

HeaderDecorator 提供的打印功能比被装饰的对象的打印功能更强大，它除了调用被装饰的对象的 print()方法，还调用自己的私有方法 printHeader()方法，打印出发票的头部。

FooterDecorator 也是一样，除了调用被装饰的对象的 print()方法，还调用自己的私有方法 printFooter()，打印出发票的尾部。

为了节省篇幅，在设计中仅给出了一个头部装饰类和一个尾部装饰类，但是读者可以根据自己的需要，加入更多的头部装饰类和尾部装饰类，并动态地选择头部装饰类和尾部装饰类的各种组合，这样才能充分发挥装饰模式的优势。

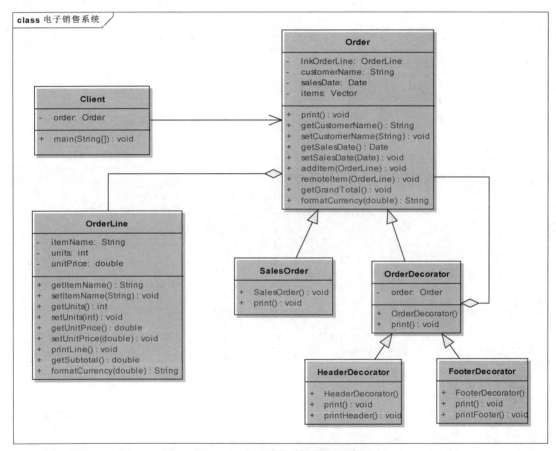

图 3-18　电子销售系统的设计方案

电子销售系统的源代码如下：

```java
package com.装饰模式.电子销售系统;
import Java.text.NumberFormat;
public class OrderLine{
    private String itemName;
    private int units;
    private double unitPrice;
    public String getItemName(){
        return itemName;
    }
    public void setItemName(String itemName){
```

```java
            this.itemName = itemName;
    }

    public int getUnits(){
        return units;
    }
    public void setUnits(int units){
        this.units = units;
    }
    public double getUnitPrice(){
        return unitPrice;
    }
    public void setUnitPrice(double unitPrice){
        this.unitPrice = unitPrice;
    }
    public void printLine(){
         System.out.println(itemName + "\t" + units
            + "\t" + formatCurrency(unitPrice)
            + "\t" + formatCurrency(getSubtotal()));
    }
    public double getSubtotal(){
        return unitPrice * units;
    }
    private String formatCurrency(double amnt){
        return NumberFormat.getCurrencyInstance().format(amnt);
    }
}
```

抽象构件（Component）角色：

```java
package com.装饰模式.电子销售系统;
import Java.util.Date;
import Java.util.Vector;
import Java.text.NumberFormat;

abstract public class Order{
    /**
     * @link aggregation
     * @directed
     * @supplierCardinality 0..*
     * @clientCardinality 1
     */
    private OrderLine lnkOrderLine;
    protected String customerName;
```

```java
    protected Date salesDate;
    protected Vector items = new Vector(10);

    public void print(){
        for (int i = 0 ; i < items.size() ; i++){
            OrderLine item = (OrderLine) items.get(i);
            item.printLine();
        }
    }
    public String getCustomerName(){
        return customerName;
    }
    public void setCustomerName(String customerName){
        this.customerName = customerName;
    }
    public Date getSalesDate(){
        return salesDate;
    }
    public void setSalesDate(Date salesDate){
        this.salesDate = salesDate;
    }
    public void addItem(OrderLine item){
        items.add(item);
    }
    public void remoteItem(OrderLine item){
        items.remove(item);
    }
    public double getGrandTotal(){
        double amnt = 0.0D;
         for (int i = 0 ; i < items.size() ; i++){
            OrderLine item = (OrderLine) items.get(i);
            amnt += item.getSubtotal();
        }
        return amnt;
    }
    protected String formatCurrency(double amnt){
        return NumberFormat.getCurrencyInstance().format(amnt);
    }
}
```

SalesOrder 代表发票的主体部分：

```java
package com.装饰模式.电子销售系统;
public class SalesOrder extends Order{
```

```java
    public SalesOrder() {
    }
    public void print(){
        super.print();
    }
}
```

抽象装饰（Decorator）角色：

```java
package com.装饰模式.电子销售系统;
abstract public class OrderDecorator extends Order{
    protected Order order;
    public OrderDecorator(Order order){
        this.order = order;
        this.setSalesDate( order.getSalesDate() );
        this.setCustomerName( order.getCustomerName() );
    }
    public void print(){
        super.print();
    }
}
```

具体装饰（ConcreteDecorator）角色1：

```java
package com.装饰模式.电子销售系统;
public class HeaderDecorator extends OrderDecorator{
    public HeaderDecorator(Order anOrder){
        super(anOrder);
    }
    public void print(){
        this.printHeader();
        super.order.print();
    }
    private void printHeader(){
        System.out.println("\t***\tIN V O I C E\t***\nXYZ Incorporated\nDate of Sale: " + order.getSalesDate());
        System.out.println("=================================================");
        System.out.println("Item\t\tUnits\tUnit Price\tSubtotal");
    }
}
```

具体装饰（ConcreteDecorator）角色2：

```java
package com.装饰模式.电子销售系统;
public class FooterDecorator extends OrderDecorator {
    public FooterDecorator(Order anOrder){
```

```
            super(anOrder);
        }
        public void print(){
            super.order.print();
            printFooter();
        }
        private void printFooter(){
            System.out.println("=======================================================");
            System.out.println("Total\t\t\t\t" +
                formatCurrency(super.order.getGrandTotal()));
        }
    }
```

测试程序如下：

```
package com.装饰模式.电子销售系统;
import Java.util.Date;
public class Client {
    /**
     * @directed
     */
    private static Order order;
    public static void main(String[] args) {
        order = new SalesOrder();
        order.setSalesDate(new Date());
        order.setCustomerName("XYZ Repair Shop");
        OrderLine line1 = new OrderLine();
        line1.setItemName("FireWheel Tire");
        line1.setUnitPrice(154.23);
        line1.setUnits(4);
        order.addItem(line1);
        OrderLine line2 = new OrderLine();
        line2.setItemName("Front Fender");
        line2.setUnitPrice(300.45);
        line2.setUnits(1);
        order.addItem(line2);
        order = new HeaderDecorator(new FooterDecorator(order));
        order.print();
    }
}
```

程序运行结果如图 3-19 所示。

```
<terminated> Client (11) [Java Application] C:\Java\Genuitec\Common\binary\c
            ***    I N V O I C E    ***
XYZ Incorporated
Date of Sale: Sun Aug 01 19:06:18 CST 2021
================================================
Item              Units    Unit Price    Subtotal
FireWheel Tire     4       ¥154.23       ¥616.92
Front Fender       1       ¥300.45       ¥300.45
================================================
Total                                    ¥917.37
```

图 3-19 电子销售系统 Java 实现程序运行结果

2. 点餐系统点饮料构件 Python 实现

```python
class 饮料():
    name = ""
price = 0.0
type = "饮料"
def getPrice(self):
    return self.price
def setPrice(self, price):
    self.price = price
def getName(self):
    return self.name

class 可乐(饮料):
    def __init__(self):
        self.name = "可乐"
        self.price = 4.0

class milk(饮料):
    def __init__(self):
        self.name = "milk"
        self.price = 5.0

# 除了基本配置，快餐店在卖可乐时，可以选择加冰，如果需要加冰，就要加 0.3 元
# 在卖牛奶时，可以选择加糖，如果需要加糖，就要加 0.5 元。怎么解决这样的问题
# 可以选择装饰模式解决这一类问题。
# 定义装饰类
class drinkDecorator():
    def getName(self):
        pass
    def getPrice(self):
        pass

class iceDecorator(drinkDecorator):
    def __init__(self, 饮料):
        self.饮料 = 饮料
    def getName(self):
```

```
return self.饮料.getName() + " +ice"
def getPrice(self):
return self.饮料.getPrice() + 0.3

class sugarDecorator(drinkDecorator):
def __init__(self, 饮料):
self.饮料 = 饮料
def getName(self):
return self.饮料.getName() + " +sugar"
def getPrice(self):
return self.饮料.getPrice() + 0.5

#构建好装饰类后，在具体的业务场景中，就可以与饮料类进行关联。以可乐加冰为例，示例业务场景如下
if __name__=="__main__":
  可乐_cola=可乐()
print("Name:%s"%可乐_cola.getName())
print("Price:%s"%可乐_cola.getPrice())
  ice_可乐=iceDecorator(可乐_cola)
print("Name:%s"% ice_可乐.getName())
print("Price:%s"% ice_可乐.getPrice())
```

程序运行结果如图 3-20 所示。

图 3-20 点餐系统点饮料构件 Python 实现程序运行结果

3.4 门面模式

门面模式（Facade Pattern）是对象的结构型模式，外部与一个子系统的通信必须通过一个统一的门面对象来进行。门面模式提供一个高层次的接口，使子系统更容易被使用。每一个子系统只归属于一个门面类，而且此门面类只有一个实例，也就是说它是一个单例模式，但整个系统可以有多个门面类。

一般而言，门面模式是为了降低子系统之间、客户端与实现层之间的依赖性。在构建一个层次化的系统时，也可以使用门面模式定义系统中每一层的入口，从而简化层与层之间的依赖关系。

➡️ **应用实例：非专业人士拍照**

我有一个很高大上的专业相机，喜欢自己手动调光圈、快门，这样照出来的照片才专业。

但非专业人士不懂这些，教半天他们也不会。幸好相机有门面模式，把相机调整到自动挡，只要对准目标按快门就可以了，一切由相机自动调整，不再需要手动调光圈、快门。这样，非专业人士也可以用这个相机拍照片了。

3.4.1 什么是门面模式

门面模式要求一个子系统的外部与其内部的通信必须通过一个统一的门面对象来进行。

就如同相机自动挡一样，门面模式的门面类将拍照人与专业相机内部的复杂性分隔开，使客户端只需要与门面对象（自动挡）打交道，而不需要与子系统内部的很多对象打交道。

3.4.2 门面模式的结构

门面模式没有一个一般化的类图描述，一个门面模式的示意性结构如图 3-21 所示。

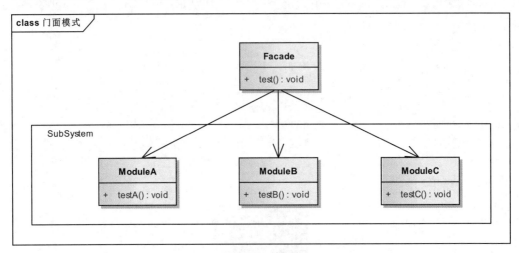

图 3-21 门面模式的示意性结构

这个结构由门面角色和子系统角色组成。

（1）门面（Facade）角色：客户端可以调用这个角色的方法。此角色知道相关的（一个或者多个）子系统的功能和责任。在正常情况下，门面角色会将所有从客户端发来的请求委派给相应的子系统。

（2）子系统（Subsystem）角色：可以有一个或者多个子系统。每一个子系统都不是单独的类，而是一个类的集合。每一个子系统都可以被客户端直接调用，或者被门面角色调用。子系统并不知道门面角色的存在，对子系统而言，门面角色仅仅是另外一个客户端而已。

3.4.3 门面模式在实际开发中的应用场景

1. 解决易用性问题

门面模式可以用来封装系统的底层实现，隐藏系统的复杂性，提供一组更加简单易用、更高层的接口。门面模式有些类似于迪米特法则（最少知识原则）和接口隔离原则：两个有交互的系统，只暴露有限的必要的接口。

2. 解决性能问题

将多个接口调用替换为一个门面接口调用，可以减少网络通信成本，提高 App 客户端的响应速度。

如果门面接口不多,可以将门面接口与非门面接口放在一起,也不需要特殊标记,当作普通接口使用即可。如果门面接口很多,我们可以在已有的接口之上,再抽象出一层,专门放置门面接口,从类、包的命名上跟原来的接口层做区分。如果门面接口特别多,并且很多都是跨多个子系统的,我们可以将门面接口放到一个新的子系统中。

3. 解决分布式事务问题

要支持两个接口调用在一个事务中执行,是比较难实现的,这涉及分布式事务的问题。虽然我们可以通过引入分布式事务框架或者事后补偿的机制解决这个问题,但代码实现都比较复杂。

现在我们可以借鉴门面模式的思想,设计一个包裹这两个操作的新接口,让新接口在一个事务中执行两个 SQL 操作。

4. 门面模式的实现

示意性代码如下。

门面(Facade)角色:

```java
package com.facade;
public class Facade {
    public void test(){
        ModuleA a=new ModuleA();
        a.testA();
        ModuleB b=new ModuleB();
        b.testB();
        ModuleC c=new ModuleC();
        c.testC();
    }
}
```

子系统(Subsystem)角色 ModuleA:

```java
package com.facade;

public class ModuleA {
    public void testA(){
        System.out.println("调用 ModuleA 中的 testA 方法");
    }
}
```

子系统(Subsystem)角色 ModuleB:

```java
package com.facade;

public class ModuleB {
    public void testB(){
        System.out.println("调用 ModuleB 中的 testB 方法");
    }
}
```

子系统（Subsystem）角色 ModuleC：

```java
package com.facade;

public class ModuleC {
    public void testC(){
        System.out.println("调用ModuleC中的testC方法");
    }
}
```

客户端：

```java
package com.facade;
public class Client {
    public static void main(String[] args) {
        Facade facade=new Facade();
        facade.test();
    }
}
```

程序运行如图 3-22 所示。

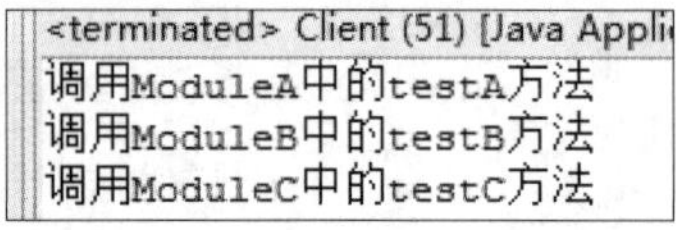

3.4.4 门面模式进阶

1．一个系统可以有几个门面类

图 3-22 门面模式程序运行结果

在门面模式中，通常只需要一个门面类，并且此门面类只有一个实例，换言之，它是一个单例类。当然，这并不意味着在整个系统里只能有一个门面类，仅仅是说每一个子系统只有一个门面类。或者说，如果一个系统有多个子系统，每一个子系统都可以有一个门面类，整个系统可以有多个门面类。

2．不能为子系统增加新行为

初学者往往以为继承一个门面类就可在子系统中加入新的行为，这种想法是错误的。门面模式的用意是为子系统提供一个集中化和简化的沟通管道，而不是向子系统中加入新的行为。

3．在什么情况下使用门面模式

（1）为一个复杂子系统提供一个简单接口。
（2）提高子系统的独立性。
（3）在层次化结构中，定义系统中每一层的入口。

3.4.5 练习

1．保安管理系统实例 Java 实现

我们通过一个保安管理系统的例子，来说明门面模式的功效。一个保安管理系统由两个录像机、三个电灯、一个遥感器和一个警报器组成。保安管理系统的操作人员需要经常将这些仪器启动和关闭。

1）不使用门面模式的设计

在不使用门面模式的情况下，这个保安管理系统的操作员必须直接操作所有的部件。在不

使用门面模式的情况下系统的设计方案如图 3-23 所示。

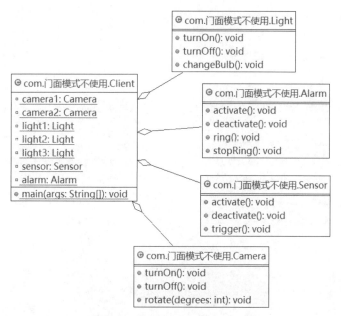

图 3-23　不使用门面模式的保安管理系统设计方案

可以看出，Client 对象需要引用所有的录像机（Camera）、电灯（Light）、感应器（Sensor）和警报器（Alarm）对象，代码如下。

录像机（Camera）：

```
package com.facadeno;
public class Camera {
    public void turnOn(){
        System.out.println("Turning on the camera.");
    }
    public void turnOff(){
        System.out.println("Turning off the camera.");
    }
    public void rotate(int degrees){
        System.out.println("rotating the camera by"+degrees+"degrees.");
    }
}
```

电灯（Light）：

```
package com.facadeno;
public class  Light {
    public void turnOn(){
        System.out.println("Turning on the light");
    }
    public void turnOff(){
System.out.println("Turning off the light");
```

```java
    }
    public void changeBulb(){
        System.out.println("changing the light-bulb.");
    }
}
```

感应器（Sensor）：

```java
package com.facadeno;
public class Sensor {
    public void activate(){
        System.out.println("Activating the sensor");
    }
    public void deactivate(){
        System.out.println("Deactivate the sensor");
    }
    public void trigger(){
        System.out.println("The sensor has been triggered.");
    }
}
```

警报器（Alarm）：

```java
package com.facadeno;
public class Alarm {
    public void activate(){
        System.out.println("Activating the alarm.");
    }
    public void deactivate(){
        System.out.println("Deactivating the alarm.");
    }
    public void ring(){
        System.out.println("Ringing the alarm.");
    }
    public void stopRing(){
        System.out.println("Stop the alarm.");
    }
}
```

客户端（Client）：

```java
package com.facadeno;
public class Client {
    static private Camera camera1 = new Camera(),camera2 = new Camera();
    static private Light light1 = new Light(),light2 = new Light(),light3 = new Light();
    static private Sensor sensor = new Sensor();
    static private Alarm alarm = new Alarm();
```

```
public static void main(String[] args) {
        camera1.turnOn();
        camera2.turnOn();
        light1.turnOn();
        light2.turnOn();
        light3.turnOn();
        sensor.activate();
        alarm.activate();
    }
}
```

程序运行结果如图 3-24 所示。

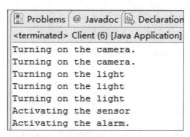

图 3-24 不使用门面模式的保安管理系统程序运行结果

2）使用门面模式的设计

一个合情合理的改进方法就是准备一个系统的控制台 SecurityFacade 作为保安管理系统的用户界面，其结构如图 3-25 所示。

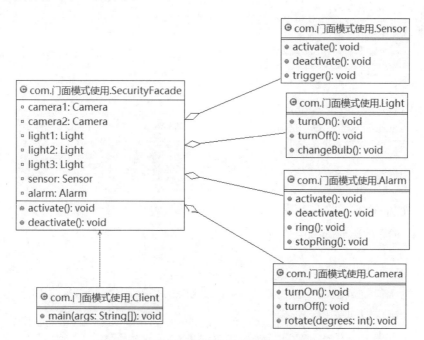

图 3-25 使用门面模式的保安管理系统的结构

录像机（Camera）、电灯（Light）、感应器（Sensor）和警报器（Alarm）代码同上，增加了一个Facade类（SecurityFacade），修改了客户端。

程序代码如下。

Facade类（SecurityFacade）：

```java
package com.facadeUse;
import com.facadeno.Alarm;
import com.facadeno.Camera;
import com.facadeno.Light;
import com.facadeno.Sensor;
public class SecurityFacade {
    private Camera camera1 = new Camera(),camera2 = new Camera();
    private Light light1 = new Light(),light2 = new Light(),light3 = new Light();
    private Sensor sensor = new Sensor();
    private Alarm alarm = new Alarm();
    public void activate(){
        camera1.turnOn();
        camera2.turnOn();
        light1.turnOn();
        light2.turnOn();
        light3.turnOn();
        sensor.activate();
        alarm.activate();
    }
    public void deactivate(){
        camera1.turnOff();
        camera2.turnOff();
        light1.turnOff();
        light2.turnOff();
        light3.turnOff();
        sensor.deactivate();
        alarm.deactivate();
    }
}
```

客户端（Client）：

```java
package com.facadeUse;
public class Client {
    public static void main(String[] args){
        SecurityFacade security = new SecurityFacade();
        security.activate();
        security.deactivate();
    }
}
```

程序运行结果与图 3-24 相同。

2. 网上点餐子系统 Python 实现

网页接收客户端的需求,并将客户端的需求发送给对应的子系统,由对应的子系统完成工作。比如说我们点了一份套餐,套餐里有一杯冰可乐,一个汉堡,一份薯条。在点餐子系统得到你的点餐后会发送信息告诉后厨需要一份套餐,于是后厨有三个人开始行动:一个做可乐,一个做汉堡,一个做薯条。这个例子里,你就是客户端,点餐子系统的网页为门面,后厨做东西的三个人为三个子系统,他们合作完成这份套餐的制作。

代码实现如下:

```
class Waiter():
    def make_set_meal_1(self):
        Coke().make()
        Hamburger().make()
        French_fries().make()


class Coke():
    def make(self):
        print('making coke')


class Hamburger():
    def make(self):
        print('making hamburger')


class French_fries():
    def make(self):
        print('making french fries')


class Client():
    def order(self):
        Waiter().make_set_meal1()


you=Client()
you.order()
```

程序运行结果如图 3-26 所示。

使用门面模式还有一个附带的好处,就是能够选择性地暴露方法。一个模块中定义的方法可以分成两部分,一部分是给子系统外部使用的,一部分是在子系统内部模块之间相互调用时使用的。

有了 Facade 类,子系统内部模块之间相互调用的方法就不用暴露给子系统外部了。

图 3-26 网上点餐子系统运行结果

第 4 章 行为型模式

行为型模式（Behavioral Pattern）是对在不同的对象之间划分责任和算法的抽象化。行为型模式不仅仅是关于类和对象的，还是关于它们之间的相互作用的。

行为型模式分为类的行为型模式和对象的行为型模式两种。

（1）类的行为型模式使用继承关系分配行为。

（2）对象的行为型模式使用对象的聚合分配行为。

本章介绍的行为型模式包括以下 5 种：策略模式、模板方法模式、命令模式、状态模式和观察者模式。

4.1 策略模式

策略模式（Strategy Pattern）属于对象的行为型模式。策略模式针对一组算法，将每一个算法封装到具有共同接口的独立的类中，从而使它们可以相互替换。策略模式可以使算法在不影响客户端的情况下发生变化。

通俗地讲，就是将那些使用的算法分别封装成独立的类，然后使用接口或抽象类将这些类统一管理起来，让需要使用这些算法的用户能够随时调用它们。

策略模式把行为和环境分开。环境类负责维持和查询行为类，具体的策略类提供各种算法。由于算法和环境已经分开，算法的增、减、修改都不会影响环境和客户端。

策略模式作为一种软件设计模式，指对象有某个行为，但是在不同的场景中，该行为有不同的实现。比如，每个人都要交个人所得税，但是在美国交个人所得税和在中国交个人所得税就有不同的计算方式。

策略模式的使用场景有以下两种。

（1）一个系统需要动态地在几种算法中选择一种。

（2）如果在一个系统里，许多类之间的区别仅在于它们的行为（方法），那么使用策略模式可以动态地让一个对象在许多行为中选择一种行为，即这些行为可互相替代。

4.1.1 应用实例：旅游出行

在旅游时，出行的方式可能是乘飞机、火车，也可能是骑自行车，设计方案如图 4-1 所示。

第 4 章 行为型模式

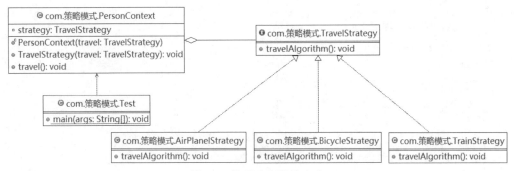

图 4-1 旅游出行设计方案

可以看出，PersonContext 扮演了环境角色。AirPlanelStrategy、TrainStrategy 和 BicycleStrategy 是具体策略角色。TravelStrategy 是抽象策略角色，它规定了具体的策略类的规范。

4.1.2 策略模式的结构

策略模式是对算法的包装，它把使用算法的责任和算法本身分割开来，委派给不同的对象管理。策略模式通常把一个系列的算法包装到一系列的策略类里面，作为一个抽象策略类的子类。用一句话来说，就是准备一组算法，并将每一个算法封装起来，使它们可以互换。下面就以一个示意性的实现讲解策略模式实例的结构，如图 4-2 所示。

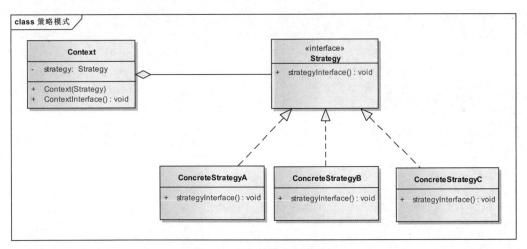

图 4-2 策略模式实例的结构

这个模式涉及 3 个角色。

（1）环境（Context）角色：持有一个 Strategy 的引用。

（2）抽象策略（Strategy）角色：这是一个抽象角色，通常由一个接口或抽象类实现。此角色给出所有的具体策略类所需的方法。

（3）具体策略（ConcreteStrategy）角色：包装了相关的算法或行为。

4.1.3 策略模式源代码

环境角色类：

```
package com.Strategy;
```

```java
public class Context {
    //持有一个具体策略的对象
    private Strategy strategy;
    /**
     * 构造函数,传入一个具体策略对象
     * @param strategy   具体策略对象
     */
    public Context(Strategy strategy){
        this.strategy = strategy;
    }
    /**
     * 策略方法
     */
    public void ContextInterface(){

        strategy.strategyInterface();
    }
}
```

抽象策略类:

```java
package com.Strategy;

public interface Strategy {
    /**
     * 策略方法
     */
    public void strategyInterface();
}
```

具体策略类1:

```java
package com.Strategy;
public class ConcreteStrategyA implements Strategy{
    public void strategyInterface() {
//相关的业务
    }
}
```

具体策略类2:

```java
package com.Strategy;

public class ConcreteStrategyB implements Strategy {
    public void strategyInterface() {
        //相关的业务
    }
}
```

4.1.4 认识策略模式

1．策略模式的重心

策略模式的重心不是如何实现算法，而是如何组织、调用这些算法，从而让程序结构更灵活，具有更好的维护性和扩展性。

2．算法的平等性

策略模式一个很大的特点就是各个策略算法是平等的。对于一系列具体的策略算法，它们的地位是完全相同的，正是因为这种平等性，才能实现算法之间的相互替换。所有的策略算法在实现上也是相互独立的，相互之间是没有依赖的。

所以，可以这样描述这一系列策略算法：策略算法是相同行为的不同实现。

3．运行时策略的唯一性

程序运行期间，虽然可以动态地在不同的策略实现中切换，但是同时只能使用一个具体的策略实现对象。

4．公有的行为

经常见到的是，所有的具体策略类都有一些公有的行为。这时候，就应当把这些公有的行为放到共同的抽象策略角色 Strategy 类里面。当然，这时抽象策略角色必须要用 Java 抽象类实现，而不使用接口实现。

这其实也是典型的继承等级结构向上方集中的标准做法，如图 4-3 所示。

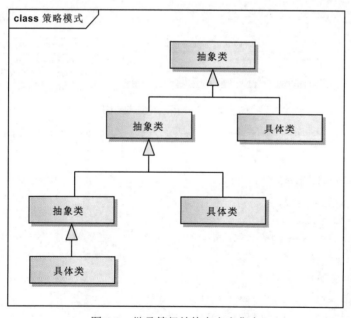

图 4-3　继承等级结构向上方集中

4.1.5 策略模式的优缺点

1．策略模式的优点

（1）策略模式提供了管理相关的算法族的办法。策略类的等级结构定义了一个算法或行为族。恰当使用继承可以把公共的代码移到父类里面，从而避免代码重复。

（2）使用策略模式可以避免使用多重条件（if...else...）语句。多重条件语句不易维护，它把采取哪一种算法或采取哪一种行为的逻辑与算法或行为的逻辑混合在一起,全部列在一个多重条件语句里面,比使用继承还要原始和落后。

2. 策略模式的缺点

(1) 策略类的数量增多。每一个策略都是一个类,复用的可能性很小,类数量增多。

(2) 所有的策略类都需要对外暴露。上层模块必须知道有哪些策略,然后才能决定使用哪一个策略。

4.1.6 排序策略系统 Java 实现

假设要设计一个排序系统(Sorter System),动态地决定采用二元排序(Binary Sort)、冒泡排序(Bubble Sort)、堆排序(Heap Sort)、快速排序(Quick Sort)、桶排序(Bucket Sort)。

显然,采用策略模式把几种排序算法包装到不同的算法类里,让所有的算法类具有相同的接口,这是一个很好的设计。排序策略系统的结构如图 4-4 所示。

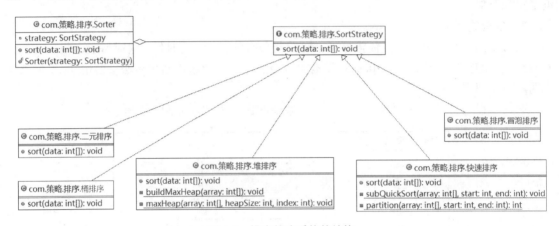

图 4-4 排序策略系统的结构

这样就可以很方便地根据需要调用不同的排序算法。排序策略系统代码实现如下。

环境角色类:

```java
package com.策略.排序;
public class Sorter {
    private SortStrategy strategy;
    public void sort(int[] data){
        strategy.sort(data);
    }
    public Sorter(SortStrategy strategy) {
        // TODO Auto-generated constructor stub
        this.strategy=strategy;
    }
}
```

抽象策略类:

```java
package com.策略.排序;
public interface SortStrategy {
```

```java
    public void sort(int [] data);
}
```

具体策略类 1：

```java
package com.策略.排序;
public class 二元排序 implements SortStrategy {
    public void sort(int[] data) {
        for (int i = 0; i < data.length; i++) {
            int temp = data[i];
            int left = 0;
            int right = i - 1;
            int mid = 0;
            while (left <= right) {
                mid = (left + right) / 2;
                if (temp < data[mid]) {
                    right = mid - 1;
                } else {
                    left = mid + 1;
                }
            }
            for (int j = i - 1; j >= left; j--) {
                data[j + 1] = data[j];
            }
            if (left != i) {
                data[left] = temp;
            }
        }
    }
}
```

具体策略类 2：

```java
package com.策略.排序;
public class 冒泡排序 implements SortStrategy {
    public void sort(int [] data) {
        // TODO Auto-generated method stub
        int temp = 0;
        for (int i = data.length - 1; i > 0; --i) {
            boolean isSort = false;
            for (int j = 0; j < i; ++j) {
                if (data[j + 1] < data[j]) {
                    temp = data[j];
                    data[j] = data[j + 1];
                    data[j + 1] = temp;
                    isSort = true;
```

```java
            }
        }
        /**
         *如果一次内循环中发生了交换，则继续比较
         *如果一次内循环中没有发生任何交换，则认为已经完成排序
         */
        if (!isSort)
            break;
    }
}
```

具体策略类3：
```java
package com.策略.排序;
public class 堆排序 implements SortStrategy {
    public void sort(int[] data) {
        // TODO Auto-generated method stub
        if (data == null || data.length <= 1) {
            return;
        }
        buildMaxHeap(data);
        for (int i = data.length - 1; i >= 1; i--) {
            ArrayUtils.exchangeElements(data, 0, i);
            maxHeap(data, i, 0);
        }
    }
    private static void buildMaxHeap(int[] array) {
        if (array == null || array.length <= 1) {
            return;
        }
        int half = array.length / 2;
        for (int i = half; i >= 0; i--) {
            maxHeap(array, array.length, i);
        }
    }
    private static void maxHeap(int[] array, int heapSize, int index) {
        int left = index * 2 + 1;
        int right = index * 2 + 2;
        int largest = index;
        if (left < heapSize && array[left] > array[index]) {
            largest = left;
        }
        if (right < heapSize && array[right] > array[largest]) {
```

```
            largest = right;
        }
        if (index != largest) {
            ArrayUtils.exchangeElements(array, index, largest);

            maxHeap(array, heapSize, largest);
        }
    }
}
```

具体策略类 4：

```
package com.策略.排序;
public class 快速排序 implements SortStrategy {
    public void sort(int[] data) {
        subQuickSort(data, 0, data.length - 1);
    }
    private static void subQuickSort(int[] array, int start, int end) {
        if (array == null || (end - start + 1) < 2) {
            return;
        }
        int part = partition(array, start, end);
        if (part == start) {
            subQuickSort(array, part + 1, end);
        } else if (part == end) {
            subQuickSort(array, start, part - 1);
        } else {
            subQuickSort(array, start, part - 1);
            subQuickSort(array, part + 1, end);
        }
    }
    private static int partition(int[] array, int start, int end) {
        int value = array[end];
        int index = start - 1;

        for (int i = start; i < end; i++) {
            if (array[i] < value) {
                index++;
                if (index != i) {
                    ArrayUtils.exchangeElements(array, index, i);
                }
            }
        }
        if ((index + 1) != end) {
```

```
            ArrayUtils.exchangeElements(array, index + 1, end);
        }
        return index + 1;
    }
}
```

具体策略类 5：
```java
package com.策略.排序;
import Java.util.ArrayList;
import Java.util.List;
public class 桶排序 implements SortStrategy {
    @SuppressWarnings({ "rawtypes", "unchecked" })
    public void sort(int[] data) {
        // TODO Auto-generated method stub
        // 找到最大数，确定要排序几趟
        int max = 0;
        for (int i = 0; i < data.length; i++) {
            if (max < data[i]) {
                max = data[i];
            }
        }
        // 判断位数
        int times = 0;
        while (max > 0) {
            max = max / 10;
            times++;
        }
        // 建立十个队列
        List<ArrayList> queue = new ArrayList<ArrayList>();
        for (int i = 0; i < 10; i++) {
            ArrayList queue1 = new ArrayList();
            queue.add(queue1);
        }
        for (int i = 0; i < times; i++) {
            for (int j = 0; j < data.length; j++) {
                int x = data[j] % (int) Math.pow(10, i + 1)
                        / (int) Math.pow(10, i);
                ArrayList queue2 = queue.get(x);
                queue2.add(data[j]);
                queue.set(x, queue2);
            }
            // 收集
            int count = 0;
```

```
                for (int j = 0; j < 10; j++) {
                    while (queue.get(j).size() > 0) {
                        ArrayList<Integer> queue3 = queue.get(j);
                        data[count] = queue3.get(0);
                        queue3.remove(0);
                        count++;
                    }
                }
            }
        }
}
```

测试程序如下：

```
package com.策略.排序;
import Java.util.Arrays;
public class Client {
    public static void main(String[] args) {
        int[] data = { 100, 3, 6, 25 };
        int[] data2 = { 1001, 32, 63, 253 };
        int[] data3 = { 10012, 322, 635, 25355 };
        SortStrategy strategy = new 冒泡排序();
        System.out.println("冒泡排序前: " + Arrays.toString(data));
        Sorter sorter = new Sorter(strategy);
        sorter.sort(data);
        System.out.println("冒泡排序后: " + Arrays.toString(data));
        SortStrategy strategy1 = new 二元排序();
        System.out.println("二元排序前: " + Arrays.toString(data2));
        Sorter sorter1 = new Sorter(strategy1);
        sorter1.sort(data2);
        System.out.println("二元排序后: " + Arrays.toString(data2));
        SortStrategy strategy2 = new 堆排序();
        System.out.println("堆排序前: " + Arrays.toString(data3));
        Sorter sorter2 = new Sorter(strategy2);
        sorter2.sort(data3);
        System.out.println("堆排序后: " + Arrays.toString(data3));
    }
}
```

程序运行结果如图 4-5 所示。

```
<terminated> Client (1) [Java Application] 
冒泡排序前: [100, 3, 6, 25]
冒泡排序后: [3, 6, 25, 100]
二元排序前: [1001, 32, 63, 253]
二元排序后: [32, 63, 253, 1001]
堆排序前: [10012, 322, 635, 25355]
堆排序后: [322, 635, 10012, 25355]
```

图 4-5 排序策略系统程序运行结果

4.1.7 练习

1．电子商务网站购物车系统 Java 实现

1）需求分析

假设现在要设计一个销售各类书籍的电子商务网站的购物车系统，计算购物车内书籍总价的最简单的情况就是把所有货品的单价乘以数量，但是实际情况肯定比这要复杂。比如，本网站可能对所有的高级会员提供每本 20%的促销折扣，对中级会员提供每本 10%的促销折扣，对初级会员没有折扣。

2）系统设计

根据描述，折扣是根据以下的几个算法中的一个进行的。

（1）算法一：对初级会员没有折扣。

（2）算法二：对中级会员提供 10%的促销折扣。

（3）算法三：对高级会员提供 20%的促销折扣。

使用策略模式的电子商务网站购物车系统的结构如图 4-6 所示。

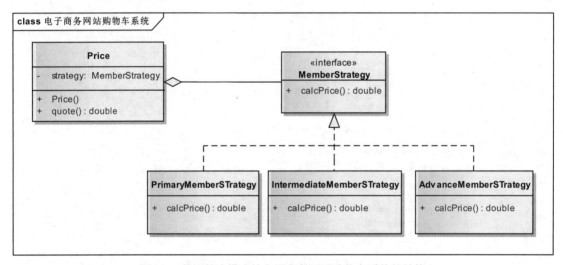

图 4-6　使用策略模式的电子商务网站购物车系统的结构

3）系统实现

源代码如下。

抽象策略（Strategy）角色：

```java
package com.策略模式.电商网站购物车;
public interface MemberStrategy {
    /**
     * 计算图书的价格
     * @param booksPrice    图书的原价
     * @return    计算出打折后的价格
     */
    public double calcPrice(double booksPrice);
}
```

具体策略（ConcreteStrategy）角色 1：

```java
package com.策略模式.电商网站购物车;
public class PrimaryMemberStrategy implements MemberStrategy {
    public double calcPrice(double booksPrice) {
        System.out.println("对于初级会员没有折扣");
        return booksPrice;
    }
}
```

具体策略（ConcreteStrategy）角色 2：

```java
package com.策略模式.电商网站购物车;
public class IntermediateMemberStrategy implements MemberStrategy {
    public double calcPrice(double booksPrice) {
        System.out.println("对于中级会员的折扣为10%");
        return booksPrice * 0.9;
    }
}
```

具体策略（ConcreteStrategy）角色 3：

```java
package com.策略模式.电商网站购物车;
public class AdvancedMemberStrategy implements MemberStrategy {
    public double calcPrice(double booksPrice) {
        System.out.println("对于高级会员的折扣为20%");
        return booksPrice * 0.8;
    }
}
```

环境（Context）角色：

```java
package com.策略模式.电商网站购物车;
public class Price {
    //持有一个具体的策略对象
    private MemberStrategy strategy;
    /**
     * 构造函数，传入一个具体的策略对象
     * @param strategy    具体的策略对象
     */
    public Price(MemberStrategy strategy){
        this.strategy = strategy;
    }
    /**
     * 计算图书的价格
     * @param booksPrice    图书的原价
     * @return    计算出打折后的价格
     */
```

```java
    public double quote(double booksPrice){
        return this.strategy.calcPrice(booksPrice);
    }
}
```

测试程序如下：

```java
package com.策略模式.电商网站购物车;
public class Client {
    public static void main(String[] args) {
        //选择并创建需要使用的策略对象
        MemberStrategy strategy = new PrimaryMemberStrategy();
        //创建环境
        Price price = new Price(strategy);
        //计算价格
        double quote = price.quote(300);
        //初级会员价格
        System.out.println("图书的初级会员价格为: " + quote);
        //高级会员价格
        strategy = new AdvancedMemberStrategy();
        price = new Price(strategy);
        quote = price.quote(300);
        System.out.println("图书的高级会员价格为: " + quote);
    }
}
```

运行结果如图4-7所示。

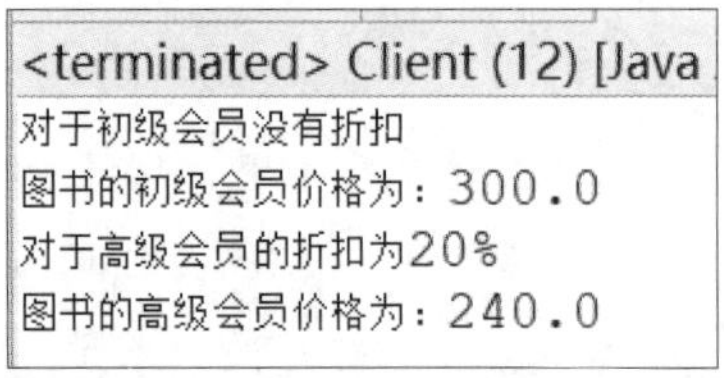

```
<terminated> Client (12) [Java
对于初级会员没有折扣
图书的初级会员价格为：300.0
对于高级会员的折扣为20%
图书的高级会员价格为：240.0
```

图4-7 电子商务网站购物车系统运行结果

从该示例可以看出，策略模式仅仅封装算法，并提供新的算法插入已有系统，让老算法从系统中"退休"。策略模式并不决定在何时使用何种算法，在什么情况下使用什么算法是由客户端决定的。

2. 商场活动 Python 实现

```python
class CashSuper(object):
    def accept_cash(self,money):
        pass
#正常收费子类
class CashNormal(CashSuper):
    def accept_cash(self,money):
        return money
```

```python
#打折收费子类
class CashRebate(CashSuper):

    def __init__(self,discount=1):
        self.discount = discount

    def accept_cash(self,money):
        return money * self.discount
#返利收费子类
class CashReturn(CashSuper):
    def __init__(self,money_condition=0,money_return=0):
        self.money_condition = money_condition
        self.money_return = money_return
    def accept_cash(self,money):
        if money>=self.money_condition:
            return money - (money / self.money_condition) * self.money_return
        return money
#具体策略类
class Context(object):
    def __init__(self,csuper):
        self.csuper = csuper
    def GetResult(self,money):
        return self.csuper.accept_cash(money)
if __name__ == '__main__':
    money = input("原价: ")
    money =float(money)
    strategy = {}
    strategy[1] = Context(CashNormal())
    strategy[2] = Context(CashRebate(0.8))
    strategy[3] = Context(CashReturn(100,10))
    mode = input("选择折扣方式(输入: 1/2或3): 1) 原价 2) 8折 3) 满100减10: ")
    mode = int(mode)
if mode in strategy.keys():
    csuper = strategy[mode]
else:
    print("不存在的折扣方式")
    csuper = strategy[1]
print("需要支付: ",csuper.GetResult(money))
```

程序运行结果如图 4-8 所示。

```
原价: 879
选择折扣方式(输入：1/2或3): 1) 原价 2) 8折 3) 满100减10: 2
需要支付: 703.2
```

图 4-8　商场活动 Python 实现运行结果

4.2　模板方法模式

模板方法模式（Template Method Pattern）是类的行为型模式。在模板方法模式中，一个抽象模板类公开定义了执行它的方法的方式或模板，该类负责给出一个算法的骨架。

模板方法模式由一个模板方法和若干个基本方法构成，它的子类可以按需要重写方法实现，但调用以抽象类中定义的方式进行。这种类型的设计模式属于行为型模式。

模式意图：定义一个操作中的算法的骨架，将一些具体操作步骤延迟到子类中。模板方法模式可以使子类在不改变一个算法的结构的情况下重定义该算法的某些特定步骤。

主要解决：一些方法通用，却在每一个子类都重新写了这一方法。

何时使用：有一些通用的方法。

如何解决：将这些通用算法抽象出来。

关键代码：在抽象类实现，其他步骤在子类实现。

> **应用实例**：西天取经八十一难

唐僧在西天取经途中经历的九九八十一难都是菩萨控制的项目，下面以模板方法模式的观点加以分析，以助理解。

唐僧一共经历了九九八十一难，而这八十一难都是在五方揭谛等暗中引导下经历的。换言之，菩萨在事先制定了一个顶级逻辑框架，而将逻辑的细节留给具体子类（在这里就是"唐僧"类）去实现，"八十一难"由西天取经方法按顺序一一调用。

按照模板方法模式的观点，具体子类可以是"唐僧"，也可以是"宋僧""明僧""阿拉伯僧"等，只要去取经，都要经历"八十一难"。

菩萨设定的诸僧取经的模板方法模式解析如图 4-9 所示。

由此可见，菩萨当年就是使用模板方法模式处理唐僧去西天取经时的项目安排的。在这个模拟系统中，关键性的角色就是抽象类"八十一难"，具体子类"唐僧"类只是按照事先安排好的步骤一步步实现而已。

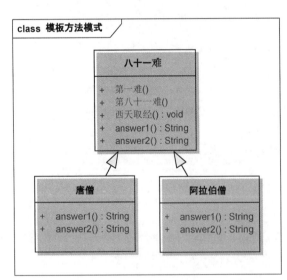

图 4-9　西天取经八十一难

4.2.1 模板方法模式的结构

模板方法模式涉及两个角色,模板方法模式的静态结构如图4-10所示。

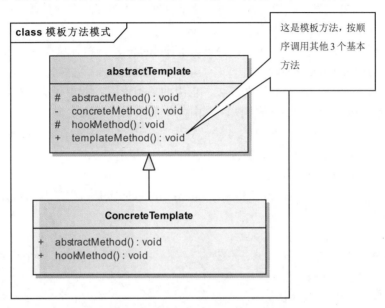

图 4-10　模板方法模式的静态结构

在图4-10的模板方法模式结构图中,抽象模板角色有以下责任。

(1)定义一个或多个基本方法,这些基本方法由抽象方法、具体方法、钩子方法组成。

(2)定义并实现一个模板方法。这个模板方法一般是一个具体方法,它给出了一个顶级逻辑的骨架,或者说它定义了算法的骨架,并按某种顺序调用其他基本方法,而抽象模板角色中其他相应的抽象方法推迟到子类实现。顶级逻辑也有可能调用一些具体方法。

具体模板角色有以下责任。

(1)实现父类定义的一个或多个抽象方法,它们是一个顶级逻辑的组成步骤。

(2)每一个抽象模板角色都可以有任意多个具体模板角色与之对应,而每一个具体模板角色都可以给出这些抽象方法(也就是顶级逻辑的组成步骤)的不同实现,从而使顶级逻辑的实现各不相同。

4.2.2 模板方法模式中的方法

模板方法中的方法可以分为两大类:模板方法和基本方法。

1. 模板方法

模板方法定义了算法的骨架,按某种顺序调用基本方法。

一个抽象类可以有任意多个模板方法。每一个模板方法都可以调用任意多个具体方法。

2. 基本方法

基本方法是整个算法中的一个步骤,基本方法又可以分为3种:抽象方法(Abstract Method)、具体方法(Concrete Method)和钩子方法(Hook Method)。

(1)抽象方法:抽象方法由抽象类声明,由具体子类实现。

(2)具体方法:具体方法由抽象类声明并实现,在具体子类中可以继承或重写它。

（3）钩子方法：在抽象类中已经平庸实现，包括用于判断的逻辑方法和需要子类重写的空方法两种。

钩子方法的作用：如果只想用某个接口中的其中一个方法，就可以编写一个抽象类实现这个接口，在这个抽象类里将钩子方法设置为平庸实现，其他方法进行具体业务实现，然后继承这个抽象类，从而使子类不需要实现其他方法。

在图4-10模板方法模式结构图中，AbstractTemplate是一个抽象类，它带有三个基本方法和一个模板方法templateMethod()。

（1）abstractMethod()是一个抽象方法，它由抽象类声明，并由子类实现。
（2）hookMethod()是一个钩子方法，它由抽象类声明并平庸实现。
（3）concreteMethod()是一个具体方法，它由抽象类声明并具体业务实现。
（4）templateMethod()负责按顺序调用上述的三个基本方法。

3. 模板方法模式Java源代码

抽象模板角色类中的abstractMethod()、hookMethod()等基本方法是顶级逻辑的组成步骤，这个顶级逻辑由templateMethod()方法代表。

```java
package com.TemplateMethod;
public abstract class AbstractTemplate {
    /**
     * 模板方法
     */
    public void templateMethod(){
        //调用基本方法
        abstractMethod();
        hookMethod();
        concreteMethod();
    }
    /**
     * 基本方法的声明（由子类实现）
     */
    protected abstract void abstractMethod();
    /**
     * 基本方法（空方法）
     */
    protected void hookMethod(){}
    /**
     * 基本方法（已经实现）
     */
    private final void concreteMethod(){
        //业务相关的代码
    }
}
```

具体模板角色类实现了父类声明的基本方法，abstractMethod()方法代表的是强制子类实现的剩余逻辑，而hookMethod()方法是可选择实现的逻辑，不是必须实现的。

```
package com.TemplateMethod;
public class ConcreteTemplate extends AbstractTemplate{
    //基本方法的实现
    public void abstractMethod() {
    //业务相关的代码
    }
    //重写父类的钩子方法
    public void hookMethod() {
    //业务相关的代码
    }
}
```

模板方法模式的关键：子类可以置换父类的可变部分，但是子类不可以改变模板方法代表的顶级逻辑。

每当定义一个新的子类时，不要按照控制流程的思路去想，而应当按照"责任"的思路去想。换言之，应当考虑哪些操作是必须置换的，哪些操作是可以置换的，以及哪些操作是不可以置换的。使用模板模式可以使这些责任变得清晰。

4.2.3 "西天取经八十一难"Java 实现

"西天取经八十一难"用模板方法模式实现的代码如下。

抽象模板角色类"八十一难"：

```
package com.TemplateMethod.TM;
public abstract class 八十一难 {
    public final void 第一难(){
    System.out.println("受难："+answer1());
    }
    public abstract String answer1();
    public abstract String answer2();
    public final void 第二难(){
        System.out.println("受难："+answer2());
    }
    public final void 西天取经(){
        第一难();
        第二难();
    }
}
```

具体模板角色类"唐僧"：

```
package com.TemplateMethod.TM;
public class 唐僧 extends 八十一难{
    public String answer1() {
        return "金蝉遭贬";
    }
```

```java
    public String answer2() {
        return "琵琶洞受苦";
    }
}
```

具体模板角色类"阿拉伯僧":

```java
package com.TemplateMethod.TM;
public class 阿拉伯僧 extends 八十一难{
    public String answer1() {
        return "孤岛奇遇";
    }
    public String answer2() {
        return "王子历险记";
    }
}
```

客户端:

```java
package com.TemplateMethod.TM;
public class Client {
    public static void main(String[] args) {
        八十一难 a = new 唐僧();
        System.out.println("唐僧的经历");
        a.西天取经();
        System.out.println("========");
        System.out.println("阿拉伯僧的经历");
        八十一难 b = new 阿拉伯僧();
        b.西天取经();
    }
}
```

程序运行结果如图 4-11 所示。

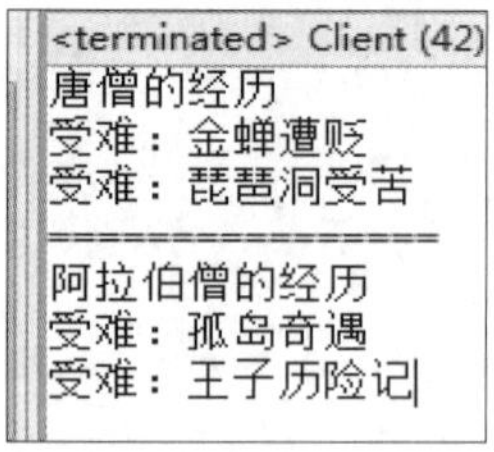

```
<terminated> Client (42)
唐僧的经历
受难：金蝉遭贬
受难：琵琶洞受苦
================
阿拉伯僧的经历
受难：孤岛奇遇
受难：王子历险记
```

图 4-11 "西天取经八十一难"Java 实现程序运行结果

4.2.4 模板方法模式进阶

通过研究 Spring 的 HibernateTemplate 源代码，可以发现 HibernateTemplate 也是借助了模板方法模式的思想。在《Design Patterns：Elements of Reusable Object-Oriented Software》，即《设计模式：可复用的面向对象软件元素》一书中，对于模板方法模式 GOF 给出了用抽象类来实现的方法，而 Spring 却把它抽象到接口的层次，使其具有更高的扩展性，这是一种创新的做法。

这种做法的思路是把变化的行为抽象成一个接口,模板改为具体类,因为已经不需要继承,所以接口就作为参数传入模板的方法,具体的变化者如"唐僧""阿拉伯僧"等只需要实现接口(变化的部分)即可。这时的结构如图4-12所示。

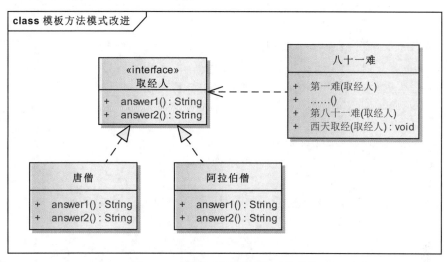

图4-12 模板方法模式改进后的结构

源代码如下。

取经人接口:

```
package com.TemplateMethod.进阶;
public interface 取经人 {
    String answer1();
    String answer2();
}
```

接口实现类"唐僧":

```
package com.TemplateMethod.进阶;
public class 唐僧 implements 取经人{
    public String answer1() {
        return "金蝉遭贬";
    }
    public String answer2() {
        return "琵琶洞受苦";
    }
}
```

接口实现类"阿拉伯僧":

```
package com.TemplateMethod.进阶;
public class 阿拉伯僧 implements 取经人{
    public String answer1() {
        return "孤岛奇遇";
    }
```

```java
    public String answer2() {
        return "王子历险记";
    }
}
```

模板类：

```java
package com.TemplateMethod.进阶;
public class 八十一难 {
    public void 第一难(取经人 a){
        System.out.println("受难: "+a.answer1());
    }
    public void 第二难(取经人 b){
        System.out.println("受难: "+b.answer2());
    }
    public void 西天取经(取经人 c){
        第一难(c); 第二难(c);
    }
}
```

客户端：

```java
package com.TemplateMethod.进阶;
public class Client {
    public static void main(String[] args) {
        八十一难 sufferer = new 八十一难();
        取经人 a = new 唐僧();
        System.out.println("唐僧的经历");
        sufferer.西天取经(a);
        System.out.println("================");
        System.out.println("阿拉伯僧的经历");
        取经人 b = new 阿拉伯僧();
        sufferer.西天取经(b);
    }
}
```

程序运行结果与图 4-11 所示的结果相同。

4.2.5 练习

1. 存款利息计算系统 Java 实现

1）系统需求分析

假设系统需要支持两种存款账户，即 Money Market 账户和 Certificate of Deposit 账户。这两种账户的存款利息是不同的，因此，在计算一个账户的存款利息时，必须区分两种不同的账户类型。

这个系统的总行为应当是计算利息，这也就决定了模板方法模式的顶级逻辑应当是利息的计算。利息计算涉及两个步骤，一个基本方法给出账户类型，另一个基本方法给出利息百分比，这两个基本方法构成具体逻辑。因为账户的类型不同，所以具体逻辑会有所不同。

2）系统设计

系统需要一个抽象角色给出顶级行为的实现，两个作为细节步骤的基本方法留给具体子类实现。需要考虑的账户有 Money Market 账户和 Certificate of Deposite 账户两种。系统的类结构如图 4-13 所示。

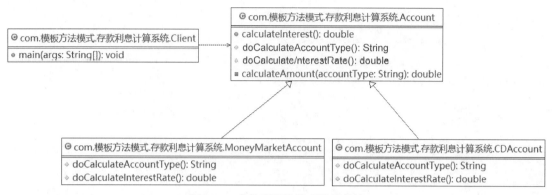

图 4-13　存款利息计算系统的结构

3）源代码

抽象模板角色：

```java
package com.模板方法模式.存款利息计算系统;
public abstract class Account {
    /**
     * 模板方法，计算利息数额
     * @return    返回利息数额
     */
    public final double calculateInterest(){
        double interestRate = doCalculateInterestRate();
        String accountType = doCalculateAccountType();
        double amount = calculateAmount(accountType);
        return amount * interestRate;
    }
    /**
     * 基本方法留给子类实现
     */
    protected abstract String doCalculateAccountType();
    /**
     * 基本方法留给子类实现
     */
    protected abstract double doCalculateInterestRate();
    /**
     * 基本方法，已经实现
     */
    private double calculateAmount(String accountType){
```

```java
        /**
         * 省略相关的业务逻辑
         */
        return 7243.00;
    }
}
```

具体模板角色1：

```java
package com.模板方法模式.存款利息计算系统;

public class CDAccount extends Account {

    @Override
    protected String doCalculateAccountType() {
        return "Certificate of Deposite";
    }

    @Override
    protected double doCalculateInterestRate() {
        return 0.06;
    }
}
```

具体模板角色2：

```java
package com.模板方法模式.存款利息计算系统;
public class MoneyMarketAccount extends Account {
    @Override
    protected String doCalculateAccountType() {

        return "Money Market";
    }
    @Override
    protected double doCalculateInterestRate() {
        return 0.045;
    }
}
```

测试程序如下：

```java
package com.模板方法模式.存款利息计算系统;
public class Client {
    public static void main(String[] args) {
        Account account = new MoneyMarketAccount();
        System.out.println("Money Market 账户的利息数额为：" + account.calculateInterest());
        account = new CDAccount();
```

```
            System.out.println("Certificate of Deposite 的利息数额为: " + account.calculateInterest());
        }
    }
```

程序运行结果如图 4-14 所示。

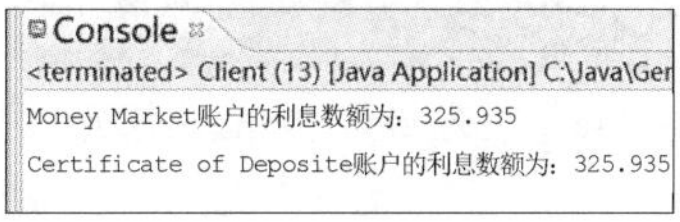

```
Console
<terminated> Client (13) [Java Application] C:\Java\Ger
Money Market账户的利息数额为: 325.935
Certificate of Deposite账户的利息数额为: 325.935
```

图 4-14 存款利息计算系统程序运行结果

2. 客户点单处理流程 Python 实现

客户点单后，系统应完成打印小票、制作结果的信息显示。

```python
#现金收费抽象类，实现一个客户点单后的处理流程流程
class User:
    def __init__(self, name, shop, times, number):
        self.name = name
        self.shop = shop
        self.times = times
        self.number = number
class Handle:

    def __init__(self, user=None):
        self.user = user
    def Invoicen(self):
        # 打印小票
        string = "打印小票:\n"\
        "客户: {}"\
        "\t 商品: {}"\
        "\t 数量: {}"\
        "\t 时间: {}".format(self.user.name, self.user.shop, self.user.number, self.user.times)
        print(string)
    def Make(self):
        # 开始制作
        print("制作完成: {} 数量: {}".format(self.user.shop, self.user.number))
    def run(self):
        self.Invoicen()
        self.Make()
if __name__ == '__main__':
  test = Handle()
  xiaoming = User("小明", "汉堡", "17:50", "5")
  test.user = xiaoming
  test.run()
```

程序运行结果如图 4-15 所示。

图 4-15 客户点单处理流程 Python 实现程序运行结果

4.3 命令模式

命令模式（Command Pattern）属于对象的行为型模式。命令模式又被称为行动模式（Action Pattern）或交易模式（Transaction Pattern）。

命令模式由客户端角色、命令角色、具体命令角色、请求者角色和接收者角色 5 个角色组成。

命令模式是对命令的封装。命令模式把发出命令的责任和执行命令的责任分割开，委派给不同的对象。

每一个命令都是一个操作，请求的一方发出请求要求执行一个操作，接收的一方收到请求并执行操作。命令模式允许请求的一方和接收的一方独立开来，使请求的一方不必知道接收请求的一方的接口，也不必知道请求是怎么被接收，以及操作是否被执行、何时被执行、怎么被执行。

命令模式允许请求的一方和接收请求的一方能够独立演化，因此具有以下的优点。
（1）新的命令很容易被加入系统。
（2）允许接收请求的一方决定是否要拒绝请求。
（3）能较容易地设计一个命令队列。
（4）可以容易地实现对请求的撤销和恢复。
（5）在需要的情况下，可以较容易地将命令记入日志。

4.3.1 命令模式的结构

命令模式的结构如图 4-16 所示。

命令模式涉及 5 个角色。

（1）客户端（Client）角色：创建一个具体命令（ConcreteCommand）对象并确定其接收者。

（2）命令（Command）角色：声明一个给所有具体命令类的抽象类或接口。

（3）具体命令（ConcreteCommand）角色：定义一个接收者和行为之间的弱耦合，实现execute()方法，负责调用接收者的相应操作。execute()方法通常称为执行方法。

（4）请求者（Invoker）角色：负责调用命令对象执行请求，相关的方法称为行动方法。

（5）接收者（Receiver）角色：负责具体实施和执行一个请求。任何一个类都可以成为接收者，实施和执行请求的方法称为行动方法。

第 4 章 行为型模式

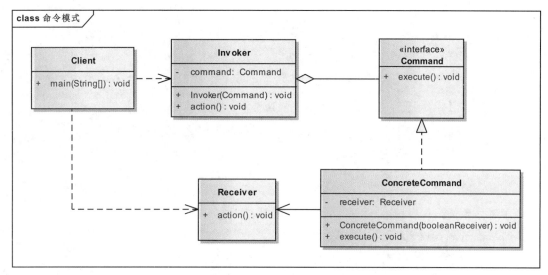

图 4-16 命令模式的结构

命令模式源代码如下。

接收者角色类：

```java
package com.command;
public class Receiver {
    /**
     * 真正执行命令相应的操作
     */
    public void action(){
        System.out.println("执行操作");
    }
}
```

抽象命令角色类：

```java
package com.command;
public interface Command {
    /**
     * 执行方法
     */
    void execute();
}
```

具体命令角色类：

```java
package com.command;

public class ConcreteCommand implements Command {
    //持有相应的接收者对象
    private Receiver receiver = null;
    /**
     * 构造方法
```

```java
         */
        public ConcreteCommand(Receiver receiver){
            this.receiver = receiver;
        }
        @Override
        public void execute() {
            //通常会调用接收者对象的相应方法，让接收者真正执行方法
            receiver.action();
        }
}
```

请求者角色类：

```java
package com.command;
public class Invoker {
    /**
     * 持有命令对象
     */
    private Command command = null;
    /**
     * 构造方法
     */
    public Invoker(Command command){
        this.command = command;
    }
    /**
     * 行动方法
     */
    public void action(){
        command.execute();
    }
}
```

客户端角色类：

```java
package com.command;
public class Client {
    public static void main(String[] args) {
        //创建接收者
        Receiver receiver = new Receiver();
        //创建命令，设定它的接收者
        Command command = new ConcreteCommand(receiver);
        //创建请求者，把命令对象设置进去
        Invoker invoker = new Invoker(command);
        //执行方法
        invoker.action();
```

            }
    }

4.3.2　应用实例：玉帝宣美猴王上天

那日玉帝命令太白金星召美猴王上天,玉皇大帝采用命令模式来处理圣旨显然是比较聪明的做法。如果不采用命令模式,玉帝就要自己处理所有的操作。在这里就包括美猴王怎样上天和怎样报到的所有细节。

这个模拟系统的结构如图4-17所示。

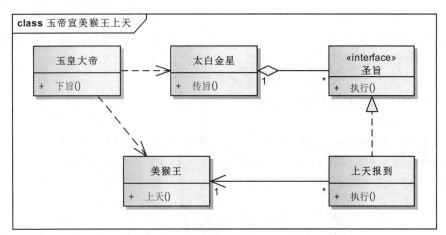

图4-17　"玉帝宣美猴王上天"模拟系统的结构

玉帝宣美猴王上天命令模式解析如下。

使用命令模式来理解玉皇大帝宣美猴王上天的过程是非常恰当的,这里面涉及5个角色。

(1)命令(Command)角色:即圣旨,这是一个抽象角色,在这里由一个Java接口实现,是所有具体圣旨必须继承的接口。

(2)具体命令(ConcreteCommand)角色:即具体圣旨,是玉皇大帝宣美猴王上天报到的那份圣旨。这个具体圣旨给出玉皇大帝和美猴王作为客户端和命令的接收者之间的弱耦合。execute()方法便是圣旨的执行方法。

(3)请求者(Invoker)角色:即太白金星。请求者角色太白金星调用此请求者角色的行动方法,要求美猴王上天。

(4)接收者(Receiver)角色:即美猴王。美猴王的上天报到的操作必须由美猴王自己做出。普天之下的万物都可以成为玉帝圣旨的接收者。

(5)客户端(Client)角色:即玉皇大帝。玉帝创建了一个具体命令对象(就是圣旨),玉帝还要指明圣旨的接收者是美猴王,并把命令交给太白金星传达。

换言之,使用了命令模式,玉皇大帝就不需要直接和美猴王打交道了,而是把打交道的细节交给太白金星处理。圣旨封装了玉皇大帝的命令,以及命令代表的操作。

4.3.3　命令模式解析

(1)请求:客户端要求系统执行的操作,在面向对象的世界里就是某个对象的方法。

(2)Command:请求封装成的对象,该对象是命令模式的主角。也就是说将请求方法封装成一个命令对象，通过操作命令对象来操作请求方法。

在命令模式中是有若干个请求的，需要将这些请求封装成一个个命令对象，客户端只需要调用不同的命令就可以达到将请求参数化的目的。

将一个个请求封装成一个个命令对象之后，客户端发起的就是一个个命令对象，而不是原来的请求方法了。

（3）Receiver：有命令，当然要有命令的接收者对象。如果只有命令，没有接受者，那不成光杆司令了？Receiver对象的主要作用就是接收命令并执行对应的操作。对使用遥控器发起的命令来说，电视机就是这个Receiver对象，比如按了待机键，电视机在收到命令后就执行了待机操作，进入待机状态。

（4）Client：有了命令对象，谁来负责创建命令？这里就引出了客户端Client对象。在命令模式中，命令是由客户端创建的。打个比方，操作遥控器的那个人，扮演的就是客户端的角色，这个人通过按下遥控器的不同按键来创建一个个命令。

（5）Invoker：现在创建命令的对象Client也已经露脸了，它负责创建一个个命令，那么谁来使用或者调度这个命令呢？命令的使用者就是Invoker对象。以人、遥控器、电视机做比喻，遥控器就是这个Invoker对象，遥控器负责使用客户端创建的命令对象。Invoker对象负责要求命令对象执行请求，通常会持有命令对象，可以持有很多的命令对象。

命令模式解析如图4-18所示。

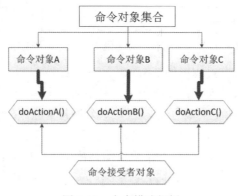

图4-18　命令模式解析

4.3.4　命令模式和策略模式的区别

1. 命令对象需要收集两类信息

命令对象要收集的信息一是方法名称，二是方法参数，通过对象Receiver收集参数。策略模式封装算法的变化，命令模式封装请求的变化。

命令模式的基本动机是解耦程序动作的发起与实际执行，这就像战场上班长下达命令（命令）：机枪掩护（命令对象1），步兵冲锋（命令对象2）。但机枪怎么打，每个步兵怎么冲锋则属于执行的事，并非都是由班长计划好的。不同的兵将会有不同的冲锋路径，这叫"多态"。

命令模式的解耦机制一方面可以实现多态，另一方面可以实现异步（动作发起并不意味着马上执行）。

2. 策略对象不需要收集方法名称信息

策略模式对象调用策略模式中某个方法的目的是实现程序行为模板，即一个程序行为中，部分动作是确定的，部分动作是不确定的，但确定部分与不确定部分有着确定的关系，比如饮料灌装流水线：第一步，准备空瓶；第二步，装满饮料；第三步，封口；第四步，贴商标。

策略模式揭示这种步骤的确定性，但具体灌什么饮料，贴什么商标则留给具体的策略实现。因此，策略模式是多态的、同步的，通常通过回调函数实现。

4.3.5 命令模式的优缺点

1．命令模式的优点

（1）更松散的耦合。命令模式使发起命令的对象——客户端和具体实现命令的对象——接收者对象完全解耦，也就是说发起命令的对象完全不知道具体实现命令的对象是谁，也不知道如何实现。

（2）更动态的控制。命令模式把请求封装起来，可以动态地对它进行参数化、队列化和日志化等，从而使系统更灵活。

（3）很自然地复合命令。在命令模式中，命令对象能够很容易地组合成复合命令，也就是宏命令，从而使系统操作更简单，功能更强大。

（4）更好的扩展性。由于发起命令的对象和具体的实现完全解耦，因此扩展新的命令很容易，只需要实现新的命令对象，然后在装配的时候，把具体的实现对象设置到命令对象中，就可以使用这个命令对象了，已有的实现完全不用变化。

2．命令模式的缺点

（1）需要很多类和对象协作，要确保正确。
（2）每个单独的命令都是一个类，增加了实现和维护的类的数量。

4.3.6 练习

1．电视机系统 Java 实现

下面就用看电视的人（Watcher），电视机（Television），遥控器（TeleController）来模拟命令模式，其中 Watcher 是 Client 角色，Television 是 Receiver 角色，TeleController 是 Invoker 角色。

设计一个简单的电视机的对象作为命令模式中的接收者。

```
package com.命令模式.电视机系统;

//电视机对象，提供了播放不同频道的方法
public class Television {

    public void playCctv1() {
        System.out.println("--CCTV1--");
    }

    public void playCctv2() {
        System.out.println("--CCTV2--");
    }
```

```java
    public void playCctv3() {
        System.out.println("--CCTV3--");
    }

    public void playCctv4() {
        System.out.println("--CCTV4--");
    }

    public void playCctv5() {
        System.out.println("--CCTV5--");
    }

    public void playCctv6() {
        System.out.println("--CCTV6--");
    }
}
```

1）非命令模式 Java 实现

不采用命令模式的系统结构如图 4-19 所示。

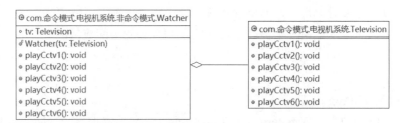

图 4-19　不采用命令模式的系统结构

```java
package com.命令模式.电视机系统.非命令模式;

import com.命令模式.电视机系统.Television;

//电视观看者类
public class Watcher {
    //持有一个
    public Television tv;

    public Watcher(Television tv) {
        this.tv = tv;
    }

    public void playCctv1() {
        tv.playCctv1();
    }
```

```java
    public void playCctv2() {
        tv.playCctv2();
    }

    public void playCctv3() {
        tv.playCctv3();
    }

    public void playCctv4() {
        tv.playCctv4();
    }

    public void playCctv5() {
        tv.playCctv5();
    }

    public void playCctv6() {
        tv.playCctv6();
    }
}
```

客户端角色类：

```java
package com.命令模式.电视机系统.非命令模式;

import com.命令模式.电视机系统.Television;

public class Client {

    /**
     * @param args
     */
    public static void main(String args[]) {
        Watcher watcher = new Watcher(new Television());
        watcher.playCctv1();
        watcher.playCctv2();
        watcher.playCctv3();
        watcher.playCctv4();
        watcher.playCctv5();
        watcher.playCctv6();
    }
}
```

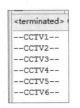

图 4-20 不采用命令模式的
电视机系统程序运行结果

运行结果如图 4-20 所示。

可以看出 Watcher 类和 Television 完全耦合了，目前本例的电视机对象只能播放 6 个电视台，如果需要添加全国所有主流卫视，需要做如下改动。

（1）修改 Television 对象，增加若干个 play××TV()方法来播放不同的卫视。

（2）修改 Watcher，也添加若干个对应的 play××TV()方法，调用 Television 的 play××TV()。

程序代码如下：

```
public void play××TV() {
    tv.play××TV();
}
```

但是这明显违背了开闭原则，是不可取的。

2）命令模式 Java 实现

采用命令模式设计的电视机系统结构如图 4-21 所示。

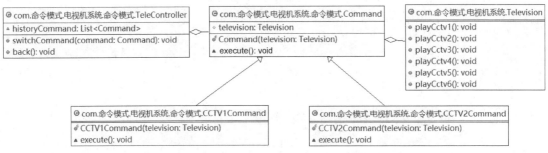

图 4-21 采用命令模式的电视机系统结构

接收者（Receiver）角色：

在本系统中，命令的接收者对象就是电视机 Television。

命令（Command）角色：

```
package com.命令模式.电视机系统.命令模式;
import com.命令模式.电视机系统.Television;
public abstract class Command {
    //命令接收者：电视机
    protected Television television;

    public Command(Television television) {
        this.television = television;
    }

    //命令执行
    abstract void execute();
}
```

具体命令（ConcreteCommand）角色：
将播放各个卫视的操作封装成一个一个命令，实现如下。
CCTV1：

```
package com.命令模式.电视机系统.命令模式;
import com.命令模式.电视机系统.Television;
//播放cctv1的命令
public class CCTV1Command extends Command {
    public CCTV1Command(Television television) {
        super(television);
        // TODO Auto-generated constructor stub
    }
    @Override
    void execute() {
        television.playCctv1();
    }
}
```

CCTV2：

```
package com.命令模式.电视机系统.命令模式;
import com.命令模式.电视机系统.Television;
//播放cctv2的命令
public class CCTV2Command extends Command {
    public CCTV2Command(Television television) {
        super(television);
        // TODO Auto-generated constructor stub
    }
    @Override
    void execute() {
        television.playCctv2();
    }
}
```

这里可增加任意个电视台。
请求者（Invoker）角色：
命令对象设计好了，那么就引入命令的调用者Invoker对象了，在此例子中，电视遥控器TeleController扮演的就是这个角色。
电视遥控器：

```
package com.命令模式.电视机系统.命令模式;

import Java.util.ArrayList;
import Java.util.List;

public class TeleController {
```

```java
//播放记录
List<Command> historyCommand = new ArrayList<Command>();

//切换卫视
public void switchCommand(Command command) {
    historyCommand.add(command);
    command.execute();
}

//遥控器返回命令
public void back() {
    if (historyCommand.isEmpty()) {
        return;
    }
    int size = historyCommand.size();
    int preIndex = size-2<=0?0:size-2;

    //获取上一个播放某卫视的命令
    Command preCommand = historyCommand.remove(preIndex);

    preCommand.execute();
}
}
```

客户端角色类：

```java
package com.命令模式.电视机系统.命令模式;

import com.命令模式.电视机系统.Television;

public class Client {
    /**
     * @param args
     */
    public static void main(String args[]) {
        //创建一个电视机
        Television tv = new Television();
        //创建一个遥控器
        TeleController teleController = new TeleController();

        teleController.switchCommand(new CCTV1Command(tv));
        teleController.switchCommand(new CCTV2Command(tv));

        teleController.switchCommand(new CCTV1Command(tv));
```

```java
        System.out.println("------返回上一个卫视--------");
        //模拟遥控器返回键
        teleController.back();
        teleController.back();
    }
}
```

程序运行结果如图 4-22 所示。

从上面的例子可以看出，命令模式的主要特点就是将请求封装成一个个命令，以命令为参数进行切换，达到请求参数化的目的。命令模式还能通过集合这个数据结构来存储已经执行的请求，进行回退操作。如果需要添加新的电视频道，只需要添加新的命令类。

在非命令模式中，看电视的人和电视耦合在一起，而命令模式则使用一个遥控器将人和电视机解耦。

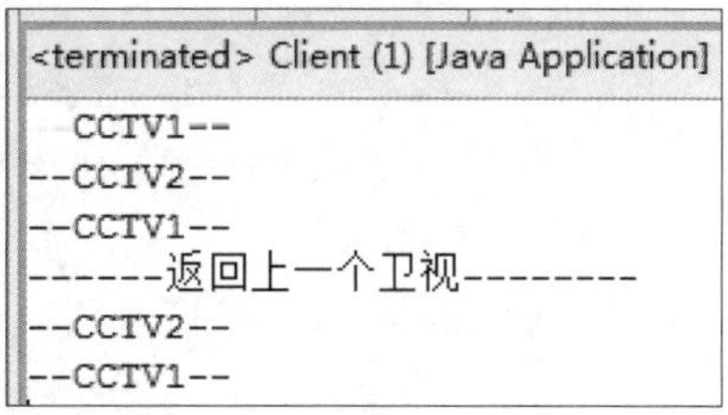

图 4-22 采用命令模式的电视机系统程序运行结果

2. 点餐系统 Python 实现

1）需求分析

在服务员已经接到顾客的点单，并录入系统后，针对不同的菜品会有不同的后台反应。比如，饭店有凉菜间、热菜间、主食间，在服务员将菜品录入系统后，凉菜间会打印出顾客点的凉菜条目，热菜间会打印出顾客点的热菜条目，主食间会打印出主食条目。这个系统的后台应该如何设计？当然，直接在场景代码中添加 if...else...判断语句是一个方法，但是这样做又一次加重了系统耦合，违反了单一职责原则，在系统需求发生变动时，又会轻易违反开闭原则，所以，我们需要重新组织一下结构。可以将该系统分为前台服务员系统和后台系统，将后台系统进一步细分为主食子系统，凉菜子系统，热菜子系统。

2）系统实现

```python
#主食子系统，凉菜子系统，热菜子系统，3个后台子系统
class backSys():
    def cook(self,dish):
        pass
class mainFoodSys(backSys):
    def cook(self,dish):
        print("MAINFOOD:Cook %s"%dish)
class coolDishSys(backSys):
    def cook(self,dish):
        print("COOLDISH:Cook %s"%dish)
class hotDishSys(backSys):
    def cook(self,dish):
        print("HOTDISH:Cook %s"%dish)


#我们可以通过命令模式实现前台服务员系统与后台系统的交互
#服务员将顾客的点单内容封装成命令，直接对后台下达命令，后台完成命令
#前台系统构建如下
```

```python
class waiterSys():
    menu_map=dict()
    commandList=[]
    def setOrder(self,command):
        print("WAITER:Add dish")
        self.commandList.append(command)

    def cancelOrder(self,command):
        print("WAITER:Cancel order...")
        self.commandList.remove(command)

    def notify(self):
        print("WAITER:Notify...")
        for command in self.commandList:
            command.execute()

#前台系统中的notify接口直接调用命令中的execute接口，执行命令。命令类构建如下
class Command():
    receiver = None
    def __init__(self, receiver):
        self.receiver = receiver
    def execute(self):
        pass
class foodCommand(Command):
    dish=""
    def __init__(self,receiver,dish):
        self.receiver=receiver
        self.dish=dish
    def execute(self):
        self.receiver.cook(self.dish)

class mainFoodCommand(foodCommand):
    pass
class coolDishCommand(foodCommand):
    pass
class hotDishCommand(foodCommand):
    pass

"""
```

Command类是父类，foodCommand类是Command的子类，相比于Command类，foodCommand类进行了一定的改造。

由于后台系统中的执行函数都是cook，因此在foodCommand类中直接将execute接口实现，如果后台系统执行的函数不同，需要在3个子命令系统中实现execute接口。

这样，后台 3 个命令类就可以直接继承，不用进行修改了。
"""

```python
#菜单类辅助业务
class menuAll:
    menu_map=dict()
    def loadMenu(self):#加载菜单，这里直接给出菜品实例
        self.menu_map["hot"] = ["Yu-Shiang Shredded Pork", "Sauteed Tofu, Home Style", "Sauteed Snow Peas"]
        self.menu_map["cool"] = ["Cucumber", "Preserved egg"]
        self.menu_map["main"] = ["Rice", "Pie"]
    def isHot(self,dish):
        if dish in self.menu_map["hot"]:
            return True
        return False
    def isCool(self,dish):
        if dish in self.menu_map["cool"]:
            return True
        return False
    def isMain(self,dish):
        if dish in self.menu_map["main"]:
            return True
        return False

#业务场景
if __name__=="__main__":
    dish_list=["Yu-Shiang Shredded Pork","Sauteed Tofu, Home Style","Cucumber","Rice"]#顾客点的菜
    waiter_sys=waiterSys()
    main_food_sys=mainFoodSys()
    cool_dish_sys=coolDishSys()
    hot_dish_sys=hotDishSys()
    menu=menuAll()
    menu.loadMenu()
    for dish in dish_list:
        if menu.isCool(dish):
            cmd=coolDishCommand(cool_dish_sys,dish)
        elif menu.isHot(dish):
            cmd=hotDishCommand(hot_dish_sys,dish)
        elif menu.isMain(dish):
            cmd=mainFoodCommand(main_food_sys,dish)
        else:
```

```
            continue
        waiter_sys.setOrder(cmd)
waiter_sys.notify()
```

程序运行结果如图 4-23 所示。

图 4-23 点餐系统 Python 实现程序运行结果

4.4 状态模式

状态模式（State Pattern）是对象的行为型模式，状态模式和对象的状态有关。

状态模式允许一个对象在其内部状态改变的时候改变其行为。例如，篮球比赛时运动员可以有正常状态、不正常状态和超常状态；曾侯乙编钟有发音和静音两种状态等。

状态模式把研究的对象的行为包装在不同的状态对象里，每一个状态对象都是一个抽象状态类的一个子类。

状态模式需要为每一个系统可能取得的状态创建一个状态类的子类。在系统的状态发生变化时，系统会改变所选的子类。

意图：允许对象在内部状态发生改变时改变它的行为，对象看起来好像修改了它的类。

主要解决：对象的行为依赖于它的状态（属性），并且可以根据它的状态改变它的相关行为。

何时使用：代码中包含大量与对象状态有关的条件语句。

如何解决：将各种具体的状态类抽象出来。

关键代码：状态模式的接口中有一个或多个方法。在状态模式中，实现类的方法一般是返回值，或者是改变实例变量的值，实现类中的方法有不同的功能。

状态模式和命令模式一样，也可以用于消除 if...else...等条件选择语句。

应用实例：曾侯乙编钟

曾侯乙编钟被誉为古代世界的"第八大奇迹"，于 1978 年在中国湖北省随县（今随州市）曾侯乙墓出土，现存于湖北省博物馆，是东周时期（战国早期）周王族诸侯国中姬姓曾国的一套大型礼乐重器。全套编钟重约 5 吨，共 65 件，分三层八组，由大小不同的青铜钟按照音调高低依次排列，每件钟均能奏出呈三度音阶的双音，悬挂在呈曲尺形的铜木结构钟架上，以敲打的方式进行演奏。

曾侯乙编钟如图 4-24 所示。

图 4-24　曾侯乙编钟

用状态模式分析，曾侯乙编钟有能够动态变化的属性，也就是它发出的声音。编钟的这种属性称为状态，而编钟也因为有这样的属性而被称为有状态的对象。在乐师不断打击各钟时，每个编钟的状态因乐师的击打而不断变化。整个编钟的行为，也就是钟所发出的声音就会从一种状态过渡到另一种状态，图 4-25 所示。

图 4-25　乐师使用编钟演奏

曾侯乙编钟系统可以用 UML 图表述，如图 4-26 所示。

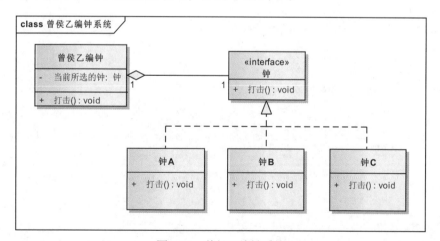

图 4-26　曾侯乙编钟系统

4.4.1　状态模式的结构

状态模式把研究的对象的行为包装在不同的状态对象里，每一个状态对象都属于一个抽象状态类的一个子类。状态模式的意图是让一个对象在其内部状态改变的时候，其行为也随之改变。状态模式的示意性结构如图 4-27 所示。

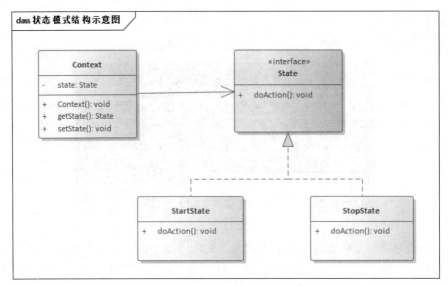

图 4-27 状态模式的示意性结构

状态模式所涉及的角色有以下 3 种。

（1）环境（Context）角色：也称上下文，定义客户端感兴趣的接口，并且保留一个具体状态类的实例。这个具体状态类的实例给出此环境对象的现有状态。

（2）抽象状态（State）角色：定义一个接口，用来封装环境（Context）对象的一个特定的状态对应的行为。

（3）具体状态（ConcreteState）角色：每一个具体状态类都实现了环境（Context）的一个状态对应的行为。

状态模式示意性源代码如下。

环境（Context）角色类：

```java
package com.状态模式;
//环境角色类
public class Context {
    //持有一个 State 类型的对象实例
    private State state;

    public Context(){
        state = null;
    }

    public void setState(State state){
        this.state = state;
    }

    public State getState(){
        return state;
    }
}
```

抽象状态（State）类：

```java
package com.状态模式;

public interface State {
  /**
   * 状态对应的处理
   */
    public void doAction(Context context);
}
```

具体状态（ConcreteState）类 StartState：

```java
package com.状态模式;

public class StartState implements State {

  public void doAction(Context context) {
    System.out.println("Player is in start state");
    context.setState(this);
  }

  public String toString(){
    return "Start State";
  }
}
```

具体状态（ConcreteState）类 StopState：

```java
package com.状态模式;

public class StopState implements State {

  public void doAction(Context context) {
    System.out.println("Player is in stop state");
    context.setState(this);
  }

  public String toString(){
    return "Stop State";
  }

}
```

客户端类：

```java
package com.state;
public class Client {
```

```java
    public static void main(String[] args){
        //创建环境
        Context context = new Context();
        //创建状态
        State startState = new StartState();
        //将状态设置到环境中
        startState.doAction(context);
        System.out.println(context.getState().toString());

        StopState stopState = new StopState();
        stopState.doAction(context);
        System.out.println(context.getState().toString());
    }
}
```

```
Player is in start state
Start State
Player is in stop state
Stop State
```

图 4-28 状态模式程序运行结果

程序运行结果如图 4-28 所示。

从上面可以看出，环境类 Context 的行为 request()是委派给某一个具体状态类的。使用多态性原则，可以动态地改变环境类 Context 的属性 State 的内容，使其从指向一个具体状态类转换为指向另一个具体状态类，从而使环境类的行为 request()由不同的具体状态类执行。

4.4.2 练习

1. 曾侯乙编钟系统 Java 实现

曾侯乙编钟系统所涉及的角色有以下 3 种。

（1）抽象状态（State）角色：此角色由"钟"接口扮演。这个接口用来规范所有的钟的行为。

（2）具体状态（ConcreteState）角色：此角色由一个个的"钟 A""钟 B""钟 C"扮演。每一个钟都有特定的发声频率。

（3）环境（Context）角色：定义"曾侯乙编钟"，并且保有一个钟的实例（即"当前所选的钟"属性），这个实例给出此编钟现在发出的声音。

源代码如下。

抽象状态（State）角色：

```java
package com.State.曾侯乙编钟;
public interface 钟 {
    /**
     * 状态对应的处理
     */
    public void doAction(String sampleParameter);
}
```

具体状态（ConcreteState）角色：

```
package com.State.曾侯乙编钟;
public class 钟A implements 钟 {
    public void doAction(String sampleParameter) {
        System.out.println("钟A : " + sampleParameter);
    }
}
```

环境（Context）角色：

```
package com.State.曾侯乙编钟;
public class 曾侯乙编钟 {
    //持有一个State类型的对象实例
    Private 钟 state;
    public void setState(钟 state) {
        this.state = state;
    }
    /**
     * 用户感兴趣的接口方法
     */
    public void request(String sampleParameter) {
        state.doAction(sampleParameter);
    }
}
```

客户端：

```
package com.State.曾侯乙编钟;
public class 乐师 {
    public static void main(String[] args){
        //创建状态
        钟 state = new 钟B();
        play(state);
        state = new 钟A();
        play(state);
        state = new 钟C();
        play(state);
    }
    public static void play(钟 state){
        //创建环境
        曾侯乙编钟 context = new 曾侯乙编钟();
        //将状态设置到环境中
    context.setState(state);
    //请求
        context.request("测试");
    }
}
```

程序运行结果如图 4-29 所示。

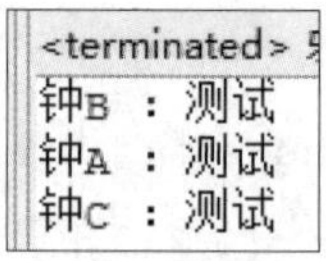

图 4-29　曾侯乙编钟系统 Java 实现程序运行结果

2. 曾侯乙编钟系统 Python 实现

```python
# 抽象的状态接口，基类
class State():
    def pronunciation(self):
        pass
    def mute(self):
        pass
# 继承于状态接口的具体状态类
class State_钟一(State):
    def pronunciation(self):
        print('状态一正常：哆')
    def mute(self):
        print('状态一正常：静音')


class State_钟二(State):
    def pronunciation(self):
        print('状态二正常：唻')
    def mute(self):
        print('状态二正常：静音')


class State_钟三(State):
    def pronunciation(self):
        print('状态三正常：咪')
    def mute(self):
        print('状态三正常：静音')
# 环境对象
class Environment():
    state = State
    def set_state(self, state_obj):
        self.state = state_obj

    # 给环境设置不同的状态，在这个不同的状态下，调用同一种方法
    # 但是得到的响应是不同的，就好像改变了类一样
    def porform(self):
        self.state.pronunciation()
        self.state.mute()
```

```python
# 测试
if __name__ == '__main__':
    e = Environment()
    select_state = State_钟一()
    e.set_state(select_state)
    e.porform()
    print('-------------')
    select_state = State_钟二()
    e.set_state(select_state)
    e.porform()
    print('-------------')
    select_state = State_钟三()
    e.set_state(select_state)
    e.porform()
```

程序运行结果如图 4-30 所示。

3. 在线投票系统 Java 实现

1）需求分析

在线投票系统要求同一个用户只能投一票，如果一个用户反复投票，而且投票次数超过 5 次，则判定为恶意刷票，系统要取消该用户的投票资格，同时也要取消该用户投的票。如果一个用户的投票次数超过 8 次，则将该用户拉入黑名单，禁止该用户再次登录和使用系统。

图 4-30　曾侯乙编钟系统 Python 实现程序运行结果

2）在线投票系统设计

采用状态模式设计的系统结构如图 4-31 所示。

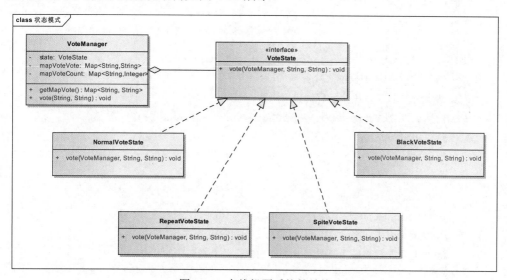

图 4-31　在线投票系统的结构

（1）投票过程分为 4 种状态。
- 正常投票。

- 反复投票。
- 恶意刷票。
- 进入黑名单。

（2）创建一个投票管理对象（相当于Context）。

（3）在线投票系统源代码如下。

```java
package com.状态模式.在线投票系统;
//环境类
import Java.util.HashMap;
import Java.util.Map;
public class VoteManager {
    //持有状体处理对象
    private VoteState state = null;
    //记录用户投票的结果,Map<String,String>对应Map<用户名称，投票的选项>
    private Map<String,String> mapVote = new HashMap<String,String>();
    //记录用户投票次数,Map<String,Integer>对应Map<用户名称，投票的次数>
    private Map<String,Integer> mapVoteCount = new HashMap<String,Integer>();
    /**
     * 获取用户投票结果的Map
     */
    public Map<String, String> getMapVote() {
        return mapVote;
    }
    /**
     * 投票
     * @param user      投票人
     * @param voteItem  投票的选项
     */
    public void vote(String user,String voteItem){
        //1.为该用户增加投票次数
        //从记录中取出该用户已有的投票次数
        Integer oldVoteCount = mapVoteCount.get(user);
        if(oldVoteCount == null){
            oldVoteCount = 0;
        }
        oldVoteCount += 1;
        mapVoteCount.put(user, oldVoteCount);
        //2.判断该用户的投票类型,就相当于判断对应的状态
        //到底是正常投票、重复投票、恶意投票还是上黑名单的状态
        if(oldVoteCount == 1){
            state = new NormalVoteState();
        }
        else if(oldVoteCount > 1 && oldVoteCount < 5){
```

```java
            state = new RepeatVoteState();
        }
        else if(oldVoteCount >= 5 && oldVoteCount <8){
            state = new SpiteVoteState();
        }
        else if(oldVoteCount > 8){
            state = new BlackVoteState();
        }
        //然后调用状态对象来进行相应的操作
        state.vote(user, voteItem, this);
    }
}
package com.状态模式.在线投票系统;
//抽象状态类
public interface VoteState {
    /**
     * 处理状态对应的行为
     * @param user        投票人
     * @param voteItem    投票项
     * @param voteManager   投票上下文，用来在实现状态对应的功能处理的时候，回调上下文的数据
     */
    public void vote(String user,String voteItem,VoteManager voteManager);
}
package com.状态模式.在线投票系统;
//具体状态类——正常投票
public class NormalVoteState implements VoteState {

    public void vote(String user, String voteItem, VoteManager voteManager) {
        //正常投票，记录到投票记录中
        voteManager.getMapVote().put(user, voteItem);
        System.out.println("恭喜投票成功");
    }

}
package com.状态模式.在线投票系统;
//具体状态类——重复投票
public class RepeatVoteState implements VoteState {

    public void vote(String user, String voteItem, VoteManager voteManager) {
        // TODO Auto-generated method stub
        //重复投票，暂时不做处理
        System.out.println("请不要重复投票");
    }
```

```java
}
package com.状态模式.在线投票系统;
//具体状态类——恶意刷票
public class SpiteVoteState implements VoteState {

    public void vote(String user, String voteItem, VoteManager voteManager) {
        // 恶意投票,取消用户的投票资格,并取消投票记录
        String str = voteManager.getMapVote().get(user);
        if(str != null){
            voteManager.getMapVote().remove(user);
        }
        System.out.println("你有恶意刷屏行为,取消投票资格");
    }
}
package com.状态模式.在线投票系统;
//具体状态类——黑名单
public class BlackVoteState implements VoteState {

    public void vote(String user, String voteItem, VoteManager voteManager) {
        //记录在黑名单中,禁止登录系统
        System.out.println("进入黑名单,禁止登录和使用本系统");
    }
}
```

测试程序如下:

```java
package com.状态模式.在线投票系统;
public class Client {
    public static void main(String[] args) {
        int j = 0;
        VoteManager vm = new VoteManager();
        for(int i=0;i<11;i++){
            j = i+1;
            System.out.print("投标次数: " +j + " 次。 ");
            vm.vote("u1","A");
        }
    }
}
```

运行结果如图 4-32 所示。

从上面的示例可以看出,状态的转换基本上是内部行为,主要在状态模式内部维护。比如,对于投票的人员,任何时候他的操作都是投票,但是投票管理对象的处理却不一定一样,投票管理对象会根据投票的次数来判断状态,然后根据状态选择不同的处理方式。

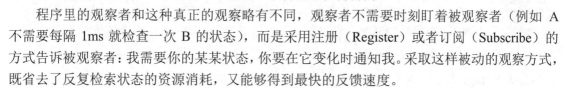

图 4-32　在线投票系统 Java 实现程序运行结果

4.5　观察者模式

观察者模式（Observer Pattern）属于行为型模式。当对象间存在一对多关系时，使用观察者模式就可以在一个对象被修改时自动通知依赖于它的对象。

观察者模式面向的需求是：A 对象（观察者）对 B 对象（被观察者）的某种变化高度敏感，需要在 B 对象发生变化后做出反应。例如，警察抓小偷，警察需要在小偷伸手作案的时候实施抓捕。在这个例子里，警察是观察者，小偷是被观察者，警察需要时刻盯着小偷的一举一动，才能保证不会错过任何瞬间。

程序里的观察者和这种真正的观察略有不同，观察者不需要时刻盯着被观察者（例如 A 不需要每隔 1ms 就检查一次 B 的状态），而是采用注册（Register）或者订阅（Subscribe）的方式告诉被观察者：我需要你的某某状态，你要在它变化时通知我。采取这样被动的观察方式，既省去了反复检索状态的资源消耗，又能够得到最快的反馈速度。

在实际中，一个软件系统常常有以下要求。

（1）在某一个对象的状态发生变化时，某些其他对象做出相应的改变。为了使系统易于复用，应选择低耦合的设计方案。

（2）降低对象之间的耦合有利于系统的复用，但是同时需要使这些低耦合的对象之间能够维持行动的协调一致，保证高度的协作。

观察者模式是满足上述要求的设计方案中最重要的一种。

4.5.1　观察者模式结构

观察者模式所涉及的角色有以下几种。

（1）抽象主题（Subject）角色：抽象主题角色又称为抽象被观察者（Observable）角色，它把对观察者对象的引用保存在一个聚集（比如 ArrayList 对象）里，每个主题都可以有任何数量的观察者。抽象主题提供一个接口，可以增加或删除观察者对象。

（2）具体主题（ConcreteSubject）角色：具体主题角色又称为具体被观察者（ConcreteObservable）角色，它把有关状态存入具体观察者对象，并在具体主题的内部状态改变时，给所有登记过的观察者发出通知。

（3）抽象观察者（Observer）角色：为所有具体观察者定义的一个接口，在得到主题的通知时更新自己，这个接口也称为更新接口。

（4）具体观察者（ConcreteObserver）角色：存储与主题的状态自恰的状态。具体观察者角色实现抽象观察者角色要求的更新接口,以便使本身的状态与主题的状态相协调。如果需要，具体观察者角色可以保持一个指向具体主题对象的引用。

4.5.2 观察者模式模型

观察者模式又分为推模型和拉模型两种方式。

在推模型中，主题对象向观察者推送主题的详细信息，不管观察者是否需要。

在拉模型中，主题对象在通知观察者的时候传递的是主题对象，如果观察者需要更具体的信息，由观察者主动到主题对象中获取。

两种模型的结构分别如图 4-33 和图 4-34 所示。

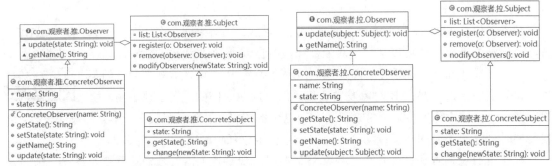

图 4-33　推模型的结构　　　　　　图 4-34　拉模型的结构

对比上面两张图可以看出，两种模型中观察者（Observer）传入的参数不同，推模型传入的参数为 String，拉模型传入的参数为 Subject。主题角色（Subject）传出的参数也不同，推模型传送的是具体内容 String，拉模型传送的是主题对象 Subject（下面代码中是 this）。具体实现代码参见下文。

1．推模型 Java 实现

抽象主题角色类：

```java
package com.观察者.推;
import Java.util.ArrayList;
import Java.util.List;
/**
 * 抽象主题角色类
 */
public abstract class Subject {
    /**
     * 保存观察者的容器
     */
    private List<Observer> list = new ArrayList<Observer>();
    /**
     * 注册观察者
```

```java
     */
    public void register(Observer o) {
        list.add(o);
        System.out.println("增加了一个观察者:" + o.getName());
    }
    /**
     * 移除观察者
     * @param observe
     */
    public void remove(Observer observe) {
        System.out.println("移除了一个观察者:" + observe.getName());
        list.remove(observe);
    }
    /**
     * 通知观察者
     * @param newState
     */
    public void nodifyObservers(String newState) {
        for (Observer observer : list) {
            observer.update(newState);
        }
    }
}
```

具体主题角色类：

```java
package com.观察者.推;
/**
 * 具体主题角色
 */
public class ConcreteSubject extends Subject {
    /**
     * 状态变化实现
     */
    private String state;

    public String getState() {
        return state;
    }

    public void change(String newState) {
        state = newState;
        System.out.println("状态变为: " + newState);
        System.out.println("开始通知观察者...");
```

```
        this.nodifyObservers(state);
    }
}
```

抽象观察者角色类:
```
package com.观察者.推;
public interface Observer {
/**
 * 更新接口
 * @param state   更新的状态
 */
    public void update(String state);
    String getName();
}
```

具体观察者角色类:
```
package com.观察者.推;
/**
 * 具体观察者角色
 */
public class ConcreteObserver implements Observer {

    private String name;
    private String state;

    public ConcreteObserver(String name) {
        this.name = name;
    }

    public String getState() {
        return state;
    }

    public void setState(String state) {
        this.state = state;
    }

    public String getName() {
        return name;
    }

    public void update(String state) {
        // 更新观察着状态
        this.state = state;
```

```
            System.out.println(getName() + "观察者状态更新为: " + state);
    }
}
```

客户端类：

```
package com.观察者.推;
public class Client {
    public static void main(String[] args) {
        Observer o1 = new ConcreteObserver("o1");
        Observer o2 = new ConcreteObserver("o2");
        Observer o3 = new ConcreteObserver("o3");
        ConcreteSubject csj = new ConcreteSubject();

        csj.register(o1);
        csj.register(o2);
        csj.register(o3);
        csj.remove(o2);
        csj.change("new State! ");    }
}
```

程序运行结果如图 4-35 所示。

```
<terminated> Test (10) [Java Application]
增加了一个观察者:o1
增加了一个观察者:o2
增加了一个观察者:o3
移除了一个观察者:o2
状态变为: new State!
开始通知观察者...
o1观察者状态更新为: new State!
o3观察者状态更新为: new State!
```

图 4-35　推模型 Java 实现程序运行结果

在程序运行时，这个客户端先创建了具体主题类的实例，以及一个观察者对象。然后，它调用主题对象的 attach()方法，将这个观察者对象向主题对象登记，也就是将它加入主题对象的聚集。同时可以看出，新添加对象的内存地址也是新的。

这时，客户端调用主题的 change()方法，改变了主题对象的内部状态。主题对象在状态发生变化时，调用超类的 notifyObservers()方法，通知所有登记过的观察者对象。

2．拉模型 Java 实现

拉模型通常把主题对象当作参数传递。具体示例代码如下。

1）抽象观察者角色类

```
package com.观察者.拉;
public interface Observer {
    /**
     * 传入主题，获取中的对象
     * @param subject
     */
```

```java
    void update(Subject subject);

    String getName();
}
```

2）具体观察者角色类

```java
package com.观察者.拉;
public class ConcreteObserver implements Observer {
    private String name;
    private String state;
    public ConcreteObserver(String name) {
        this.name = name;
    }
    public String getState() {
        return state;
    }
    public void setState(String state) {
        this.state = state;
    }
    public String getName() {
        return name;
    }
    public void update(Subject subject) {
        // 主动从主题中获取数据
        state = ((ConcreteSubject) subject).getState();
        System.out.println(getName() + "观察者状态更新为: " + state);
    }
}
```

3）抽象主题角色类

拉模型的抽象主题类主要的改变是 nodifyObservers()方法。在循环通知观察者的时候，也就是循环调用观察者的 update()方法的时候，传入的参数不同了。

```java
package com.观察者.拉;
import Java.util.ArrayList;
import Java.util.List;
public abstract class Subject {
    /**
     * 保存观察者的容器
     */
    private List<Observer> list = new ArrayList<Observer>();
    /**
     * 注册观察者
     */
    public void register(Observer o) {
```

```java
        list.add(o);
        System.out.println("增加了一个观察者:" + o.getName());
    }
    /**
     * 移除观察者
     *
     * @param o
     */
    public void remove(Observer o) {
        System.out.println("移除了一个观察者:" + o.getName());
        list.remove(o);
    }
    /**
     * 通知观察者
     *
     * @param newState
     */
    public void nodifyObservers() {
        for (Observer observer : list) {
            observer.update(this);
        }
    }
}
```

4）具体主题类

跟推模型相比，具体主题类的变化就是在调用通知观察者的方法的时候不需要传入参数了。

```java
package com.观察者.拉;
public class ConcreteSubject extends Subject {
    /**
     * 状态变化实现
     */
    private String state;
    public String getState() {
        return state;
    }
    public void change(String newState) {
        state = newState;
        System.out.println("状态变为: " + newState);
        System.out.println("开始通知观察者...");
        this.nodifyObservers();
    }
}
```

5）客户端类

同推模型。

程序运行结果同推模型。

4.5.3 两种模式的比较

推模型是假定主题对象知道观察者需要的数据。而拉模型中的主题对象不知道观察者具体需要什么数据，在没有办法的情况下，直接将自身传递给观察者，让观察者自己按需要取值。

推模型可能会使观察者对象难以复用，因为观察者的update()方法是按需要定义的参数，可能无法兼顾所有可能的使用情况。这就意味着在出现新情况的时候，可能提供新的update()方法，或者干脆重新实现观察者。拉模型就不会造成这样的情况，因为在拉模型下，update()方法的参数是主题对象本身，这基本上是主题对象能传递的最大数据集合了，基本上可以适应各种情况的需要。

4.5.4 练习

1. 观察者模式在 Java 中的应用

Java 语言的 Java.util 库提供了一个 Observable 类和一个 Observer 接口，这是 Java 语言对观察者模式的支持。Observable 类和 Observer 接口采用的是观察者模式的拉模式。观察者模式在 Java 中的应用如图 4-36 所示。

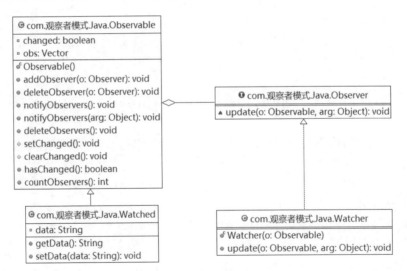

图 4-36 观察者模式在 Java 中的应用

此例具有很大的参考价值，源代码如下。

抽象主题（Subject）角色：

```java
package com.观察者模式.Java;
//抽象主题角色（抽象被观察者角色）
import Java.util.Vector;
public class Observable {
    private boolean changed = false;
    private Vector obs;
```

```java
/** Construct an Observable with zero Observers. */
public Observable() {
    obs = new Vector();
}
/**
 * 将一个观察者添加到观察者聚集
 */
public synchronized void addObserver(Observer o) {
    if (o == null)
        throw new NullPointerException();
    if (!obs.contains(o)) {
        obs.addElement(o);
    }
}
/**
 * 将一个观察者从观察者聚集中删除
 */
public synchronized void deleteObserver(Observer o) {
    obs.removeElement(o);
}
public void notifyObservers() {
    notifyObservers(null);
}
/**
 * 如果本对象有变化（那时 hasChanged() 方法会返回 true）
 * 则调用本方法通知所有登记的观察者，即调用它们的 update()方法
 * 传入 this 和 arg 作为参数
 */
public void notifyObservers(Object arg) {
    Object[] arrLocal;
    synchronized (this) {
        if (!changed)
        return;
        arrLocal = obs.toArray();
        clearChanged();
    }
    for (int i = arrLocal.length-1; i>=0; i--)
        ((Observer)arrLocal[i]).update(this, arg);
}
/**
 * 将观察者聚集清空
 */
public synchronized void deleteObservers() {
```

```java
            obs.removeAllElements();
    }
    /**
     * 将"已变化"设置为true
     */
    protected synchronized void setChanged() {
        changed = true;
    }
    /**
     * 将"已变化"重置为false
     */
    protected synchronized void clearChanged() {
        changed = false;
    }
    /**
     * 检测本对象是否已变化
     */
    public synchronized boolean hasChanged() {
        return changed;
    }
    /**
     * Returns the number of observers of this <tt>Observable</tt> object.
     * @return  the number of observers of this object.
     */
    public synchronized int countObservers() {
        return obs.size();
    }
}
```

具体主题（ConcreteSubject）角色：

```java
package com.观察者模式.Java;
//具体主题角色（具体被观察者角色）
public class Watched extends Observable{
    private String data = "";
    public String getData() {
        return data;
    }
    public void setData(String data) {
        if(!this.data.equals(data)){
            this.data = data;
            setChanged();
        }
        notifyObservers();
```

        }
    }

抽象观察者（Observer）角色：

```java
package com.观察者模式.Java;
//抽象观察者角色
public interface Observer {
    void update(Observable o, Object arg);
}
```

具体观察者（ConcreteObserver）角色：

```java
package com.观察者模式.Java;
//具体观察者角色
public class Watcher implements Observer{
    public Watcher(Observable o) {
        o.addObserver(this);
        // TODO Auto-generated constructor stub
    }
    public void update(Observable o, Object arg) {

        System.out.println("状态改变为： " + ((Watched)o).getData());
    }
}
```

测试程序如下：

```java
package com.观察者模式.Java;
public class Client {
    public static void main(String[] args) {
        //创建被观察者对象
        Watched watched = new Watched();
        //创建观察者对象，并将被观察者对象登记
        Observer watcher = new Watcher(watched);
        //给被观察者状态赋值
        watched.setData("start");
        watched.setData("run");
        watched.setData("stop");
    }
}
```

程序运行结果如图 4-37 所示。

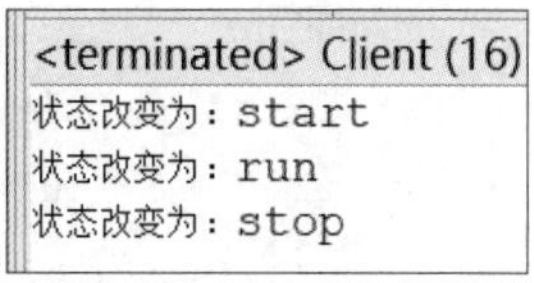

2. 新闻社推送新闻 Java 实现

新闻社需要向所有新闻订阅者推送新闻。经过需求分析我们不难发现，一家新闻社一般对应的是多个新闻订阅者。那么我们先从新闻社开始入手，分析一下新闻社应具备基本功能。

（1）新闻。

图 4-37 观察者模式在 Java 中的应用程序运行结果

（2）查看有哪些订阅者。
（3）添加订阅者。
（4）取消订阅者。
（5）推送新闻。

我们可以分别采用推模型和拉模型实现新闻社向所有新闻订阅者推送新闻的功能。推模型和拉模型的结构分别如图 4-38 和图 4-39 所示，图中标出了它们的区别。

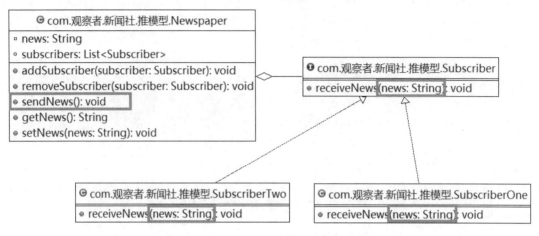

图 4-38　新闻社向所有新闻订阅者推送新闻推模型的结构

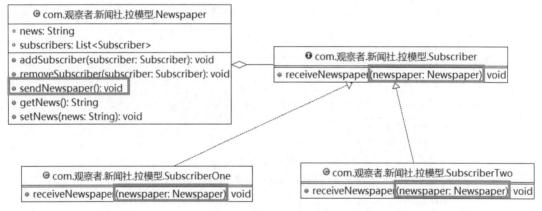

图 4-39　新闻社向所有新闻订阅者推送新闻拉模型的结构

1）推模型

```java
package com.观察者.新闻社.推模型;
import Java.util.ArrayList;
import Java.util.List;
//新闻社
public class Newspaper {
    private String news;//新闻
    private List<Subscriber>subscribers = new ArrayList();//存储所有订阅者
    public void addSubscriber(Subscriber subscriber){//添加订阅者
        subscribers.add(subscriber);
```

```
    }
    public void removeSubscriber(Subscriber subscriber){//删除订阅者
        subscribers.remove(subscriber);
    }
    public void sendNews(){//给所有订阅者发送新闻
        for (Subscriber subscriber : subscribers) {
            subscriber.receiveNews(news);
        }
    }
    public String getNews() {
        Return news;
    }
    public void setNews(String news) {
        this.news = news;
    }
}
```

新闻社已经规划好，现在考虑订阅者，订阅者有多人，我们先定义一套订阅者的标准（接口），并考虑订阅者需要做的事。

接收新闻：

```
package com.观察者.新闻社.推模型;
//订阅者
Public interface Subscriber {
    void receiveNews(String news);//接收新闻
}
```

标准写好了，现在有两个客户联系我们，要成为我们的订阅者：

```
package com.观察者.新闻社.推模型;
//订阅者One
public class SubscriberOne implements Subscriber{
    @Override
    public void receiveNews(String news) {
        System.out.println("订阅者One,接收新闻:"+news);
    }
}
package com.观察者.新闻社.推模型;
//订阅者Two
public class SubscriberTwo implements Subscriber{
    @Override
    public void receiveNews(String news) {
        System.out.println("订阅者Two,接收新闻:"+news);
    }
}
```

规划好新闻社并与订阅者签了合同,现在来实际运行一下。

```java
package com.观察者.新闻社.推模型;
public class Test {
    public static void main(String[] args) {
        //建立真正的新闻社
        Newspaper newspaper = new Newspaper();
        //将订阅者One加入
        newspaper.addSubscriber(new SubscriberOne());
        //将订阅者Two加入
        newspaper.addSubscriber(new SubscriberTwo());
        //设置新闻
        newspaper.setNews("70周年");
        //发送新闻
        newspaper.sendNews();
    }
}
```

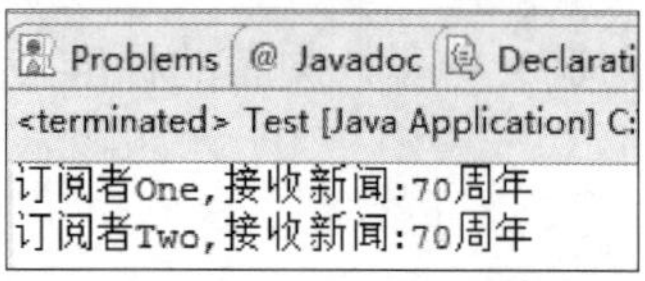

程序运行结果如图4-40所示。

2)拉模型

新闻社推送的都是新闻,订阅者订阅的是新闻的内容。我们是否可以不推送新闻,而是把新闻社提供给订阅者,让订阅者在需要时自己获取新闻呢?下面改造一下。

图4-40 新闻社向所有新闻订阅者推送新闻(推模型)程序运行结果

```java
package com.观察者.新闻社.拉模型;
import Java.util.ArrayList;
import Java.util.List;
//新闻社
public class Newspaper {
    private String news;//新闻
    public List<Subscriber>subscribers = new ArrayList();//存储所有订阅者

    public void addSubscriber(Subscriber subscriber){//添加订阅者
        subscribers.add(subscriber);
    }
    public void removeSubscriber(Subscriber subscriber){//删除订阅者
        subscribers.remove(subscriber);
    }
    //提供的不再是新闻,而是新闻社
    public void sendNewspaper(){
        for (Subscriber subscriber : subscribers) {
            subscriber.receiveNewspaper(this);//将新闻社提供给订阅者
        }
    }
    public String getNews() {
```

```java
        return news;
    }
    public void setNews(String news) {
        this.news = news;
        sendNewspaper();//在设置新闻时,将新闻社提供给订阅者
    }
}
package com.观察者.新闻社.拉模型;
//订阅者
public interface Subscriber {
    void receiveNewspaper(Newspaper newspaper);//接收新闻社
}
```

以下是两个订阅者:

```java
package com.观察者.新闻社.拉模型;
//订阅者One
public class  SubscriberOne implements Subscriber{
    @Override
    public void receiveNewspaper(Newspaper newspaper) {
        String news = newspaper.getNews();//订阅者One自行从新闻社获取新闻
        System.out.println("订阅者One,接收新闻:"+news);
    }
}
package com.观察者.新闻社.拉模型;
//订阅者two
public class  SubscriberTwo implements Subscriber{
    @Override
    public void receiveNewspaper(Newspaper newspaper) {
        String news = newspaper.getNews();//订阅者Two自行从新闻社获取新闻
        System.out.println("订阅者Two,接收新闻:"+news);
    }
}
```

测试程序如下:

```java
package com.观察者.新闻社.拉模型;
public class  Test {
    public static void main(String[] args) {
        //建立真正的新闻社
        Newspaper newspaper = new Newspaper();
        //将订阅者One加入
        newspaper.addSubscriber(new SubscriberOne());
        //将订阅者Two加入
        newspaper.addSubscriber(new SubscriberTwo());
        //设置新闻
```

```
            newspaper.setNews("70周年");
    }
}
```

运行结果与图 4-40 推模型相同。

3. 新闻社推送新闻 Python 实现（拉模型）

```python
class Newspaper:
    def __init__(self,subscribers):
        # self.news=news
        self.subscribers = []
    def __str__(self):
        return 'the {} computer'.format(self.news)
    def addSubscriber(self,subscriber):
        return self.subscribers.append(subscriber)

    def removeSubscriber(self,subscriber):
        return self.subscribers.remove(subscriber)

    def sendNewspaper(self):
        for subscriber in self.subscribers:
            subscriber.receiveNewspaper(self)

    def getNews(self):
        return self.news

    def setNews(self,news):
        self.news = news
        self.sendNewspaper()


class Subscriber:
    def receiveNewspaper(self,newspaper):
        pass


class SubscriberOne:
    def receiveNewspaper(self,Newspaper):
        news = Newspaper.getNews()
        print("订阅者One，接收新闻:" + news)


class SubscriberTwo:
    def receiveNewspaper(self,Newspaper):
        news = Newspaper.getNews()
        print("订阅者Two，接收新闻:" + news)
```

```
if __name__=="__main__":
    # 建立新闻社
    np = Newspaper([])
    # 将订阅者 One 和 Two 加入
    np.addSubscriber(SubscriberOne())
    np.addSubscriber(SubscriberTwo())
    # 设置新闻
    np.setNews("70 周年")
```

程序运行结果如图 4-41 所示。

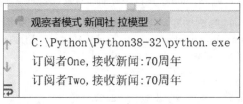

图 4-41　新闻社推送新闻 Python 实现（拉模型）程序运行结果

推模型的代码和拉模型的代码差不多，读者可以一试。

参考文献

[1] 郭双宙. 软件设计原则与模式[M]. 北京：机械工业出版社，2015.

[2] weixin_39966465. 里氏替换原则_Python 工匠：写好面向对象代码的原则（中）[EB/OL].（2020-12-16）[2021-02-12]. https://blog.csdn.net/weixin_39966465/article/details/111366101.

[3] Erich Gamma，Richard Helm，Ralph Johnson，等. 设计模式：可复用面向对象软件的基础[M]. 李英军，马晓星，蔡敏，等，译. 北京：机械工业出版社，2019.

[4] 闫宏. Java 与模式[M]. 北京：电子工业出版社，2002.

[5] 程杰. 大话设计模式[M]. 北京：清华大学出版社，2007.

[6] [美]Bruce Eckel. Java 编程思想：第 3 版[M]. 陈昊鹏，等，译. 北京：机械工业出版社，2005.